THE LAYING HEN AND ITS ENVIRONMENT

CURRENT TOPICS IN VETERINARY MEDICINE AND ANIMAL SCIENCE

VOLUME 8

Series ISBN: 90-247-2429-5

The Laying Hen and its Environment

A Seminar in the EEC Programme of Coordination of
Research on Animal Welfare, organised by R. Moss
and V. Fischbach, and held at Luxembourg,
March 11-13, 1980

Sponsored by the Commission of the European Communities,
Directorate-General for Agriculture, Coordination of
Agricultural Research

edited by

R. MOSS
Ministry of Agriculture, Fisheries and Food,
London, United Kingdom

1980

MARTINUS NIJHOFF PUBLISHERS
THE HAGUE / BOSTON / LONDON

for

THE COMMISSION OF THE EUROPEAN COMMUNITIES

Distributors

for the United States and Canada
Kluwer Boston, Inc.
190 Old Derby Street
Hingham, MA 02043
USA

for all other countries
Kluwer Academic Publishers Group
Distribution Center
P.O.Box 322
3300 AH Dordrecht
The Netherlands

ISBN-13:978-94-009-8924-5   e-ISBN-13:978-94-009-8922-1
DOI: 10.1007/978-94-009-8922-1

Publication arranged by
Commission of the European Communities,
Directorate-General Scientific and Technical Information
and Information Management

EUR 6877 EN

ⓒ ECSC, EEC, EAEC, Brussels-Luxembourg, 1980
Softcover reprint of the hardcover 1st edition 1980

Manuscript Preparation by
Janssen Services, 14 The Quay, Lower Thames Street, London, EC3R 6BU

LEGAL NOTICE

Neither the Commission of the European Communities nor any person acting on behalf of the
Commission is responsible for the use which might be made of the following information.

CONTENTS

VI

PREFACE

The welfare of domestic poultry, particularly those kept under intensive housing conditions is a subject on which many, often divergent and conflicting views, are held. This divergence, may be the result either of insufficient knowledge of the facts of particular poultry husbandry systems or a differing interpretation of those facts. With regard to poultry and the laying hen in particular, there is a need to gather together a basic knowledge of avian behaviour in all the circumstances and systems of husbandry under which birds are presently being kept. That knowledge should lead to the development of interpretative and hopefully predictive theories which in turn will allow us to meet the recommendations of Article 3 of the Council of Europe Convention on the Protection of Animals Kept for Farming Purposes: 'that animals shall be housed, provided with food, water and care in a manner which - having regard to their species and to their degree of development, adaptation and domestication - is appropriate to their physiological and ethological needs in accordance with established experience and scientific knowledge'.

In furtherance of those objectives a small group of specialists in poultry physiology and ethology, both from within and outside the European Community met by invitation of the Commission in Luxembourg between 11 and 13 March 1980.

Their objective was to discuss what is already known, and can be agreed on, of the normal physiology and ethological range of the laying hen, and secondly to try to determine what further research and development work is required to add to that knowledge.

The papers and discussions which follow clearly illustrate:

(i) the considerable complexity of the subject under discussion and of the problems inherent in deciding basic criteria;

(ii) the wide area of unknown factors which need urgent investigation.

Indeed the papers and discussions will, I am sure, provoke more questions than they answer.

Finally, may I, on behalf of the scientists and experts who were present at the Seminar thank the Commission for the opportunity to participate in this most useful meeting.

R. Moss

OPENING REMARKS

R. Moss *(UK)*

    Good morning ladies and gentlemen. May I welcome you
to Luxembourg and thank each one of you for coming to participate
in this seminar. I hope that at the end of these few days which
we will spend together we will be able to define precisely those
areas in which we have sufficient knowledge and those in which
more work needs to be done.

    This is an important meeting, an important discussion,
because the results of it will not only be published in a
Commission document but will also be used in the Report which
the Commission has been asked to put to the Council of Ministers
in the early summer.

    I will now ask Mr. Connell to say a few words.

J. Connell *(CEC)*

    Good morning ladies and gentlemen. On behalf of the
Commission, welcome to Luxembourg. As Mr. Moss has indicated,
this meeting comes at a very opportune time; I hope that in the
next few days we will all learn a great deal and, hopefully,
reach some conclusions which may help us in planning our research
ahead and certainly will help the Commission eventually in
looking towards some legislation.

    I will not take up any more of your time as I am sure
you are all eager to get on with the business of the meeting.

SESSION I

CURRENT KNOWLEDGE OF 'NORMAL' RANGE OF BEHAVIOUR AND PHYSIOLOGY

Chairman: J.P. Signoret

THE ETHOGRAM OF THE DOMESTICATED HEN

I.J.H. Duncan

Behaviour has its origin in the hereditary constitution of the animal and the impact of the environment on that given genotype, both at the time of the response and during development. In some cases the action of the genes is well buffered against environmental changes; the so-called 'fixed action patterns' which are common in avian agonistic or courtship displays are good examples of this, e.g. waltzing in the courtship of the domestic cock (Wood-Gush, 1954). In other cases, it is only the potentiality to perform a behaviour pattern which is inherited and the appearance of the pattern will depend almost entirely on the animal being subjected to a certain set of conditions during development which enable it to learn the behaviour, e.g. key-pecking in the domestic hen to obtain a food reward in a Skinner box (Duncan and Hughes, 1972). Of course these are two extreme examples; most behaviour patterns lie somewhere on the continuous scale between them.

An ethogram is a behavioural profile, a catalogue of the major fixed action patterns characteristic of a species. Therefore, the ethogram will contain these behaviour patterns which are buffered against environmental effects or at least lie towards that end of the continuum. However, this classification is not exact; some action patterns may be more 'fixed' than others.

Generally speaking, the behaviour patterns shown by wild species will be those which have evolved by natural selection to fit the animal best to its environment. In other words, they will result in the animal leaving a greater number of viable offspring than would other behaviour patterns. In the case of domestic species this is no longer true. Domestication has

R. Moss (ed.), The Laying Hen and its Environment, 5-18

exerted its influence on the behaviour of species in two major ways. First, the species responds to the type of artificial domestic environment in which it is placed. Second, genetic selection of specific strains for certain desirable character- istics leads to even greater deviations from 'normal' behaviour (Kretchmer and Fox, 1975). This means that there are many difficulties associated with describing the ethogram of a domestic species such as the fowl. For example, what sort of environment should the ethogram be recorded in? The jungles of SE Asia where the wild ancestors of the fowl evolved may be entirely unsuitable for the survival of modern hybrid strains. Also, modern husbandry systems such as battery cages and deep litter pens differ in many aspects and, as we shall hear, produce very different behavioural profiles. Moreover, which strain of fowl should be used in constructing the ethogram? Modern strains show bigger differences in behaviour than the normal range of differences in the wild ancestor. For example, some strains of broiler chicken, if given the right conditions, show full incubation and brooding behaviour and these patterns are completely lacking in some light-hybrid strains of laying chicken.

I shall describe briefly, therefore, those behaviour patterns which are commonly shown in different environments by many strains of modern hybrid laying fowl.

FEEDING BEHAVIOUR

Feeding behaviour consists of two phases, an appetitive phase and a consummatory phase. The appetitive phase contains the elements of locomotion and ground-scratching as well as activities involving the beak such as pecking, flicking, probing and beating. The consummatory phase is the picking up and swallowing of food particles. These main elements appear to occur in most modern strains and in Junglefowl (Wood-Gush, 1971; Kruijt, 1964). However, the proportions of the various elements may differ between strains. For example, the ratio of appetit- ive elements to consummatory elements is less in broilers than in layers (Masic et al., 1974; Savory, 1975). Also the two phases are not closely correlated (Masic et al., 1974). It has been suggested that pecking and ingestion in chicks have separate causal factors and the two behaviour patterns only become linked through experience (Hogan, 1971). If this is true it could explain the poor correlation between appetitive and consummatory elements and also why some observers have postulated the existence of a 'pecking drive' unconnected with feeding.

Ground-scratching seems to be automatically coupled with the performance of pecking movements and not released by any specific stimulus both in Junglefowl and domestic fowl (Kruijt, 1964; Wood-Gush, 1971). However, in my work on deprivation and frustration I have gained the impression that ground-scratching increases when the rate of ingestion is slow compared to the degree of hunger. This, of course, would usually be quite adaptive in the wild.

Food-running is another element of feeding behaviour which is seen in Junglefowl and domestic fowl, particularly in young birds (Kruijt, 1964; Baeumer, 1955; Spalding, 1873). It seems to be released by living or large food items and it usually results in the food object being torn into edible portions.

The motor patterns involved in picking up and swallowing particles of food have been investigated (Hutchinson and Taylor, 1962; Gentle et al., 1980). The bird strikes at the food particle with considerable force from a distance of 1 - 3 cm. Just before contact the bill is opened and this usually means that the lower mandible strikes the substrate first. The particle is then grasped, usually by the upper mandible being moved to the lower, and the head is drawn back in several jerking movements, which results in the food particle being thrown backwards into the pharynx. Removing one-third to one-half of the upper mandible causes a marked reduction in feeding efficiency and a temporary fall in food intake.

DRINKING BEHAVIOUR

Drinking behaviour in young chicks is not readily stimulated by a still water surface; droplets of water on a solid
surface are more powerful stimuli (Lloyd Morgan, 1896; Kruijt,
1964). However, once birds have learnt the source of water,
they seem capable of using a variety of techniques to obtain
it (Richardson, 1969). This suggests that Spalding (1873) may
have been correct in thinking that the newly hatched chick must
learn to drink.

COMFORT AND GROOMING BEHAVIOUR

The behaviour patterns included in this section are
preening, in which the feathers are manipulated with the bill,
bill-wiping, head-scratching, head-shaking, tail-wagging,
feather-raising, feather-ruffling, stretching and dust bathing.
Most of these terms are self-explanatory; feather-raising is
the short-term act of raising all the contour feathers to give
what Morris (1956) called a 'ruffled' posture and McFarland and
Baher (1968) called a 'raised' posture. This posture is never
maintained for longer than a few seconds and is followed by
preening, a feather-ruffle or the feather subsiding to a
'normal' (McFarland and Baher, 1968) or 'fluffed' (Morris,
1956) posture; feather-ruffling is a vigorous shaking of the
feathers with a rotatory movement of the body and it usually
follows feather-raising.

It should be emphasised that little is known of the
causative factors responsible for any of these behaviour
patterns.

Preening involves the arrangement, cleansing and general
maintenance of the health and structure of feathers by the
bill. The most important movements appear to be stroking and
nibbling (Williams and Strungis, 1979). Preening occurs not
only in response to obvious external stimuli, but also when
the bird is in a state of low arousal (Wood-Gush, 1959). It

has yet to be proved whether this preening occurs because the bird's attention switches to tactile stimuli which are constantly present, or because there is a gradual build up of peripheral stimulation due to an accumulation of skin and feather debris or because there is a build up in the internal motivation to preen.

Bill-wiping is more easily accounted for; it often occurs during or after feeding when foreign particles can be seen adhering to the bill. Similarly, head-shaking is often associated with drinking. Head-scratching, tail-wagging, feather-raising and feather-ruffling are often incorporated into bouts of preening and perhaps should be regarded as integral parts of a grooming system rather than independent patterns.

Kruijt (1964) described two types of stretching in Junglefowl: unilateral stretching in which the wing and leg on the same side are pushed out and down behind the bird, and bilateral stretching in which both wings are half opened and stretched upward and forward. Unilateral stretching certainly does occur in domestic fowl but there is some doubt whether bilateral stretching is found. A similar movement does appear in conflict situations and it has been suggested that it may be an incomplete form of wing-flapping which commonly occurs as a male display during courtship and agonistic encounters (Wood-Gush, 1971; Duncan, 1979a).

If given a dry, dusty substrate, dust bathing consists of initial pecking and scratching movements in the dust, squatting in the dust, movements of the feet and wings to raise dust into the ruffled plumage, rubbing the head and sides in the dust, followed by feather-ruffling and shaking the dust out of the feathers. However, all these motor patterns can be seen in the absence of dust and even in the absence of a solid substrate - for example, in cages. The causative factors have not been identified although it has been suggested that dust bathing may eliminate ectoparasites and also reduce excess surface lipids from the plumage. However, unlike Bobwhite quail, in which there is a strong connection between feather lipids and dust bathing (Borchelt and Duncan, 1974), in domestic fowl removal of the uropygial gland does not affect the amount of dust bathing shown and it also occurs when birds are free of ectoparasites (N.S. Williams, pers. comm.). Dust bathing is a highly social activity and is slightly unusual in that birds which are dust bathing communally come into physical contact with each other.

It should be stated that as well as occurring as comfort movements, most of these behavioural patterns also occur in thwarting and conflict situations where they may be regarded as 'displacement activities' (Duncan and Wood-Gush, 1972).

SLEEPING

Sleeping has not been widely investigated. There does seem to be a tendency for birds to roost off the ground in a perching position in order to sleep, and although this response may be taught to chicks by their dam (Wood-Gush et al., 1978), if often occurs without teaching.

It should be mentioned that there is a behavioural pattern called yawning or gaping but this is not usually associated with sleep and is probably a comfort movement.

REPRODUCTIVE BEHAVIOUR

The Junglefowl and domestic fowl show intricate courtship behaviour when both sexes are present in a flock (Wood-Gush, 1954). In the absence of cocks, hens seldom show any elements of courtship apart from occasional sexual crouching to other stimuli such as human beings, particularly at point-of-lay. Mounting and treading of hens by hens has been reported (Guhl, 1948) but must be regarded as much more unusual than in mammalian species such as cattle.

The next stage in the reproductive cycle is nesting behaviour and its exact form depends largely on the environment.
In a deep litter pen with nest boxes, nesting consists of
increased locomotion, the pre-laying call, nest examination,
nest entry, sitting, nest building, oviposition, standing,
cackling and the leaving of the nest (Duncan, 1979b). Feral
hens show most of the elements of nesting behaviour shown by
domestic hens in pens. However, in studies of a feral population in Australia, the cock had an important role in nest
site selection (McBride et al., 1969), whereas this was not the
case in studies made on a feral population in Scotland (Duncan
et al., 1978). Nesting of the birds in Scotland was characterised by secrecy and nest sites by concealment. Hens kept in
battery cages also show most of the elements of nesting behaviour but there are strain differences (Wood-Gush, 1972).

Nesting and laying behaviour in Junglefowl and one or two
strains of domestic fowl are followed by incubation and brooding
behaviour. However, generally speaking, these traits have been
bred out of modern hybrid laying stock and so will not be
discussed here. However, some modern hybrids do pay attention
to the egg they have just laid and also to eggs laid previously
in the nest (Wood-Gush, 1975). These are elements of behaviour
lying somewhere between nesting and incubation and should be
considered in the present debate.

SOCIAL BEHAVIOUR

Fowl are social animals which, when allowed to do so,
form a cohesive social group and communicate by means of audible
calls and visual displays. Wood-Gush (1971) has listed twelve
chick calls and twenty two adult calls. The most common calls
such as the food call, the ground predator alarm call, the pre-
laying call and the post-laying cackle are given in a wide
variety of environments. Many of the others may be bound to
very specific stimuli and so not be released under any commercial
conditions. None of them has been examined systematically.

Many of the visual displays are performed by the cock during courtship or agonistic encounters (Wood-Gush, 1956) and need not be considered further. Domestic hens show a few postures and displays which have signal value, and those will be considered in more detail. The position of the tail was considered to have important signal value in the feral population of fowl in Australia (McBride et al., 1969) and these observations need to be repeated with modern hybrid strains.

Agonistic behaviour which includes attack, escape, avoidance and submission, has been much studied since Schjelderup-Ebbe (1922) described the peck-order. Attacks include (a) threatening, in which one bird raises its head above the level of the other's head; (b) pecking, in which the comb, head, nape and neck are pecked; (c) chasing and (d) fighting, in which two birds face up to each other aiming pecks with their bills and kicks with their feet and spurs, and occasionally give some side display with waltzing. Submission involves crouching and remaining very still; avoidance and escape both involve locomotion and are self-explanatory.

It is generally assumed that a very extensive, complex, rich and 'natural' environment will put least constraints on the behaviour shown by the domestic fowl. Exactly the opposite is true of agonistic behaviour. Very few agonistic interactions were seen during many hours of observation of a feral fowl population living in a fairly rich habitat on a Scottish island (Wood-Gush et al., 1978). There is evidence that in a more 'natural' social group agonistic encounters will be reduced. For example, the presence of a dominant cock inhibits the expression of aggression among all birds within 6 m (McBride et al., 1969) and the presence of a mother hen reduces aggressive interactions among her brood (Fält, 1978). Also, aggressive interactions increase with increasing density down to a space allowance of about 800 cm^2 per bird. When hens are given less space than this, there is a sharp reduction in aggressive behaviour (Al-Rawi and Craig, 1975) with threatening declining before aggressive pecking (Banks and Allee, 1957; Hughes and Wood-Gush, 1977).

ANTI-PREDATOR BEHAVIOUR

As a prey species it is not surprising that the Junglefowl developed several anti-predator behaviour patterns and that these can still be seen in modern hybrid fowl. The main ones are alerting and running away in response to ground predator alarm calls (Wood-Gush, 1971), freezing and sometimes crouching to aerial predator alarm calls (Wood-Gush, 1971), extreme caution and indirect route when travelling to nest (Duncan et al., 1978), and running or sometimes flying away when disturbed at close quarters (Wood-Gush and Duncan, 1976). The question of whether or not domestic hens fly is an interesting one. Undoubtedly they can fly over vertical obstacles, as anyone will testify who has tried to fence hens without clipping their wing feathers. But do they fly if not fenced in? Hens from the feral population in Scotland were seen to fly on several occasions when disturbed on open ground. One hen flew about 80 m at a height of about 4 m in a steady directed 'flap-glide' flight. They also regularly flew up from branch to branch in stages to their roosting place 3 - 4 m high in a tree each night. Moreover, in the morning they often flew from their roosting place 50 - 60 m to a feeding area. In addition, at least one incubating hen each day flew down from her nest high on a hillside to a feeding area which was a distance of 100 - 150 m. However, walking and running were much commoner responses than flying, particularly to disturbance. Hens were never seen to fly if there was any cover whatsoever (Wood-Gush and Duncan, 1976; Duncan et al., 1978). It is possible that wing-flapping is an intention movement to fly and that this remains as an ambivalent response given by modern hybrids to disturbance in open ground whereas the full flying response has largely disappeared.

REFERENCES

Al-Rawi, B. and Craig, J.V., 1975. Agonistic behaviour of caged chickens
    as related to group size and area per bird. Appl. Anim. Ethol., 2,
    69-80.

Baeumer, E., 1955. Lebensart des Haushuhns. Z. Tierpsychol., 12, 387-401.

Banks, E.M. and Allee, W.C., 1957. Some relations between flock size and
    agonistic behaviour in domestic hens. Physiol. Zool., 30, 255-268.

Borchelt, P.L. and Duncan, L., 1974. Dust bathing and feather lipid in
    Bobwhite Quail (Colinus virginianus). Condor, 76, 471-472.

Duncan, I.J.H., 1979a. Can scientific research help in assessment of
    welfare? Proc. IV Rev. in Rural Sci., (in press).

Duncan, I.J.H., 1979b. Nesting behaviour - its control and expression.
    Proc. IV Rev. in Rural Sci., (in press).

Duncan, I.J.H. and Hughes, B.O., 1975. Feeding activity and egg formation
    in hens lit continuously. Br. Poult. Sci., 16, 145-155.

Duncan, I.J.H., Savory, C.J. and Wood-Gush, D.G.M., 1978. Observations on
    the reproductive behaviour of domestic fowl in the wild. Appl.
    Anim. Ethol., 4, 29-42.

Duncan, I.J.H. and Wood-Gush, D.G.M., 1972. An analysis of displacement
    preening in the domestic fowl. Anim. Behav., 20, 68-71.

Fält, B., 1978. Differences in aggressiveness between brooded and non-
    brooded domestic chicks. Appl. Anim. Ethol., 4, 211-221.

Gentle, M.J., Hughes, B.O. and Hubrecht, R.C., 1980. ·The effects of beak
    trimming on food intake, feeding behaviour and body weight in adult
    hens. Appl. Anim. Ethol., (in press).

Guhl, A.M., 1948. Unisexual mating ·in a flock of White Leghorn hens.
    Trans. Kans. Acad. Sci., 51, 107-111.

Hogan, J.A., 1971. The development of a hunger system in young chicks.
    Behaviour, 39, 128-201.

Hughes, B.O. and Wood-Gush, D.G.M., 1977. Agonistic behaviour in domestic
    hens: The influence of housing method and group size. Anim. Behav.
    25, 1056-1062.

Hutchinson, J.C.D. and Taylor, W.W., 1962. Motor co-ordination of pecking
    fowls. Anim. Behav., 10, 55-61.

Kretchmer, K.R. and Fox, M.I., 1975. Effects of domestication on animal
    behaviour. Vet. Rec., 96, 102-108.

Kruijt, J.P., 1964. Ontogeny of social behaviour in Burmese Red Junglefowl
(*Gallus gallus spadiceous*). Behaviour Suppl., 12.

Lloyd Morgan, C., 1896. The habit of drinking in young chicks. Science
3, 900.

McBride, G., Parer, I.P. and Foenander, F., 1969. The social organisation
and behaviour of the feral domestic fowl. Anim. Behav., Monogr.,
2, 125-181.

McFarland, D.J. and Baher, E., 1968. Factors affecting feather posture
in the Barbary dove. Anim. Behav., 16, 171-177.

Masic, B., Wood-Gush, D.G.M., Duncan, I.J.H., McCorquodale, C.C. and
Savory, C.J., 1974. A comparison of the feeding behaviour of young
broiler and layer males. Br. Poult. Sci., 15, 499-505.

Morris, D., 1956. The feather postures of birds and the problem of the
origin of social signals. Behaviour, 9, 75-113.

Richardson, A.R., 1969. Drinking behaviour of chickens at troughs and
nipples. Wld's Poult. Sci. J., 25, 144.

Savory, C.J., 1975. A growth study of broiler and layer chicks reared in
single-strain and mixed-strain groups. Br. Poult. Sci., 16, 315-318.

Schjelderup-Ebbe, T., 1922. Beiträge zur Social-psychologie des Haushuhns.
Z. Psychol., 88, 225-252.

Spalding, D., 1873. Instinct. MacMillan's Magazine, 27, 282-293.
Reprinted (1954). Br. J. Anim. Behav., 2, 2-11.

Williams, N.S. and Strungis, J.C., 1979. The development of grooming
behaviour in the domestic chicken (*Gallus gallus domesticus*).
Poult. Sci., 58, 469-472.

Wood-Gush, D.G.M., 1954. The courtship of the Brown Leghorn cock. Br. J.
Anim. Behav., 2, 95-102.

Wood-Gush, D.G.M., 1956. The agonistic and courtship behaviour of the
Brown Leghorn cock. Br. J. Anim. Behav., 4, 133-142.

Wood-Gush, D.G.M., 1959. Time-lapse photography: A technique for studying
diurnal rhythms. Physiol. Zool., 32, 272-283.

Wood-Gush, D.G.M., 1971. The Behaviour of the Domestic Fowl. London:
Heinemann.

Wood-Gush, D.G.M., 1972. Strain differences in response to sub-optimal
stimuli in the fowl. Anim. Behav., 20, 72-76.

16

Wood-Gush, D.G.M., 1975. Nest construction by the domestic hen: Some
    comparative and physiological considerations. In: Neural and
    Endocrine Aspects of Behaviour in Birds. (Ed. P. Wright, P.G. Caryl
    and D.M. Vowles), pp. 35-49. Amsterdam: Elsevier.
Wood-Gush, D.G.M. and Duncan, I.J.H., 1976. Some behavioural observations
    on domestic fowl in the wild. Appl. Anim. Ethol., 2, 255-260.
Wood-Gush, D.G.M., Duncan, I.J.H. and Savory, C.J., 1978. Observations
    on the social behaviour of domestic fowl in the wild. Biol. Behav.,
    3, 193-205.

DISCUSSION

I.J.H. Duncan *(UK)*

Perhaps I could start the discussion myself by asking if there is agreement amongst the experts here that missing out such things as incubation and brooding behaviour is valid, or should we be including these behaviour patterns in our discussions?

K. Vestergaard *(Denmark)*

I know that Dr. Fölsch has observed incubation behaviour in modern domestic breeds. I think he found that about 14% showed it under certain circumstances - they were free range, with bushes around them. So there are indications of incubation behaviour.

I.J.H. Duncan

Yes, I think that incubation is one of the behaviour patterns where there are large strain differences. I know there are one or two modern hybrid strains which do show evidence of incubation behaviour. On the other hand, other strains seem to be totally refractive and it is impossible to initiate incubation behaviour in any way by giving them any combination of environmental conditions, or even by injecting the hormones thought to initiate brooding behaviour.

M. Zanforlin *(Italy)*

I have been told by egg producers that some birds do not give the 'cackling' call after they have laid the eggs; they do not show incubation behaviour. It appears that this depends on the situation more than on the particular breed. If they are kept on the ground, birds of the same breed show both behaviour patterns - cackling after egg laying and then incubation behaviour after a certain number of eggs. However, if they are kept in cages with about 400 - 500 cm^2 per hen, they do not show either of these behaviour patterns. Do you have any evidence of this?

I.J.H. Duncan

I have no evidence at all for differences in cackling behaviour after egg laying. I could only suggest that under commercial conditions the timing of the cacklings often seems to be disrupted. Very often the cackling occurs quite a long time after oviposition, so it could easily be missed. I am very interested in your remarks on the connection between cackling and incubation.

M. Zanforlin

I have observed it myself - no cackling at all after egg laying, not even much later.

J.M. Faure *(France)*

I think that as far as brooding behaviour is concerned there is an interaction between genetic background and the environment. I do not think there is ever brooding behaviour in cages, even with a strain which is able to brood in good conditions. It seems that the genetic determinant for brooding is a very simple one compared with most of the behaviour pattern which have been studied. However, it varies between strains.

J.P. Signoret *(France)*

The discussion has moved to the strain problem so I think it would be better to have Dr. Faure's paper now before we go any further.

TO ADAPT THE ENVIRONMENT TO THE BIRD OR THE BIRD TO THE ENVIRONMENT?

J.M. Faure

ABSTRACT

The selection of birds for an improved adaptation to their environment necessitates that (a) adaptation is genetically determined, (b) it is possible to measure adaptiveness using a large number of animals and (c) it is worthwhile doing. The first point is illustrated by the results of a selection experiment on open-field activity (selection for high open-field activity is related to a decrease in fearfulness). The second point is illustrated by a few examples which show that it is possible to measure fear, cannibalistic tendencies or perching behaviour with very short or automated tests. The answer to the third point is that it may be worth selecting for adaptation from a welfare point of view but not necessarily from a commercial one.

Because of the lack of (a) commercial interest and (b) research on this subject, breeders may be reluctant to include this kind of approach in their breeding programmes.

It is true to say that chicken accommodation was designed with man in mind and not the bird. This was true even before commercial breeding but, at that time, there had not been many important changes in breeding methods and the birds had adapted to the housing system by 'natural selection' over thousands of generations. If the actual conditions of present day accommodation were allowed to remain stable over the next few thousand generations it is probable that the birds would again adapt. However, problems have arisen because of the many changes, e.g. from farm to industrial breeding, which have occurred over the last 40 years.

Artificial selection is more efficient than natural selection. We are thus faced with two choices - (a) to adapt the environment to fulfil the birds' needs or (b) to adapt the birds to the environment. If we choose the latter possibility we must first answer 3 questions:

1. Is it possible to select for behaviours relevant to adaptation problems?

2. Is it possible for a commercial breeder to select for these behaviours?

3. Is it ethically acceptable to select for adaptation to a very artificial environment? Fox (1978) said, "Would it be more humane to reduce and possibly eliminate the animal's need genetically by careful selective breeding? If it could be reduced to a 'vegetable' it would have few needs and few correlated rights".

In this paper I will try to answer only the first two questions; the third is a matter for the individual's conscience.

1. IS IT POSSIBLE TO SELECT FOR BEHAVIOUR?

At least three long term selection programmes have been undertaken in domestic fowl. They are for: sexual behaviour (Siegel, 1972); dominance ability (Ortman and Craig, 1968); and open-field activity (Faure and Folmer, 1975). Only the last is relevant to adaptation.

For this study a circular open area is used; it is 1 metre in diameter. The walls are of wood and wood litter is scattered on the floor. In the centre of the open area there is a hexagonal prism which reflects light from an infra-red bulb onto 6 photocells mounted on the wall of the open area. The light beams thus divide the open area into 6 sections. Each photocell is connected to a tally counter.

Each chick is individually tested; it is placed in one sector of the open area, that is, between two photocells, and is allowed 100 seconds to move. If it does not cut one of the light beams during this time the test is stopped and a maximum latency of 100 seconds is recorded; its activity and number of light beams broken would be 0. Activity refers to the total number of light beams cut whereas the number of cells indicates the number of different light beams cut. If the bird cuts a beam before the 100 s has elapsed, it is allowed one further period of 100 s (this second period starts as soon as the first light beam is cut).

We have selected for the product between total number of beams cut and the number of different beams cut because some birds may remain immobile in front of a photocell but, by movements of the head alone, e.g. pecking, they may cut the beam several times. The selection was made on a familial basis thus taking into account not only the bird phenotype, but also the mean of its siblings.

Figure 1 shows the progression, over seven generations, for the following characteristics: product, activity, number of

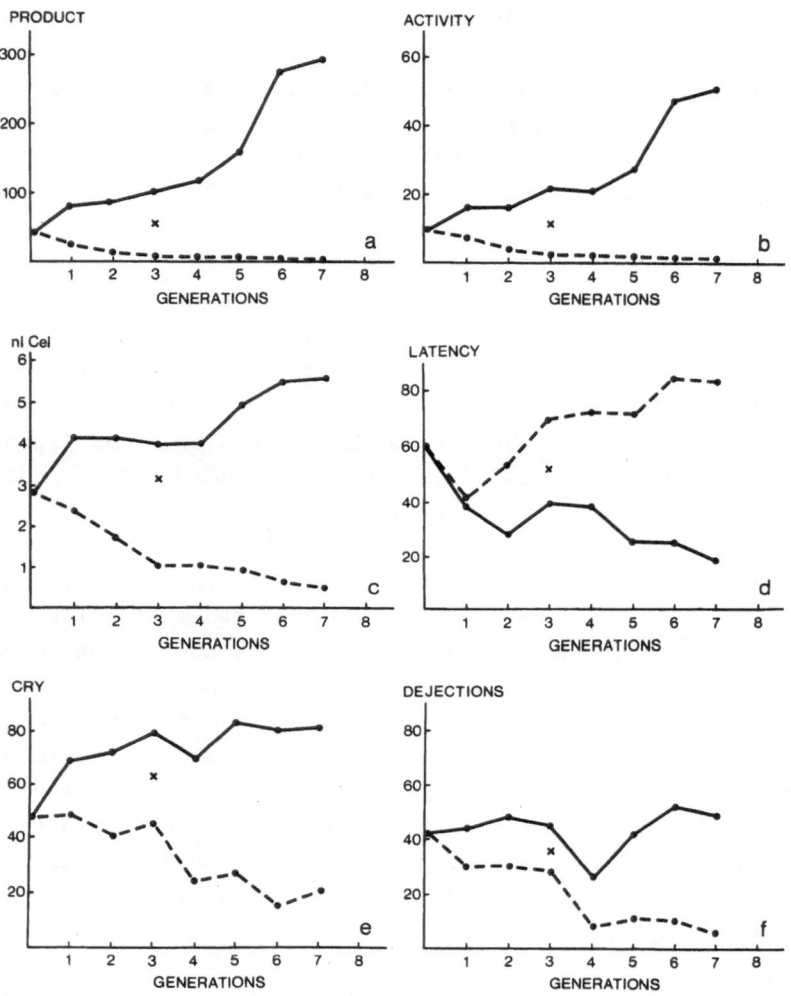

Fig. 1   Evolution, over 7 generations, of the following characteristics:
          a - product; b - activity; c - number of cells; d - latency to move;
          e - the percentage of birds which peep for more than 80% of the
          test duration; and f - the percentage of birds which defecate
          during the test. Full line: Active strain. Dotted line: inactive
          strain

cells, latency to move, percentage of birds which peep for more than 80% of the test and the percentage of birds which defecate during the test. After only two generations the differences between the strains were all highly significant and thus selection was very efficient.

Figure 2 shows the results of emotional reaction tests carried out on 2nd and 6th generation birds. The birds were classed as one of four categories from 0 (low reaction) to 3 (high reaction) following three tests of emotional reaction: (1) reaction to novel visual stimulus; (2) reaction to novel auditory stimulus; (3) duration of tonic immobility. There is a big difference between the two strains both at the 2nd and 6th generations, although this difference is more pronounced in the 6th generation birds. The active birds have low emotional reaction (fear) scores.

Table 1 shows that the active strain had a lower resting level of plasma corticosterone than did the inactive one and that the differences between the strains were exaggerated by application of a stress factor.

TABLE 1

CORTICOSTERONE LEVEL IN mg/ml FOR THE ACTIVE LINE (A) AND INACTIVE LINE (1)

Age (weeks)	2	5 Stress	6	25
A	4.49	11.00	10.04	1.24
1	6.78*	20.92***	7.80^{NS}	1.43^{NS}
A	4.21	8.10	4.86	0.53
1	5.38^{NS}	13.78**	6.46*	0.70*

*   Significant at 5% dilution
** Significant at 1% dilution
***Significant at 0.01% dilution
NS = Not significant

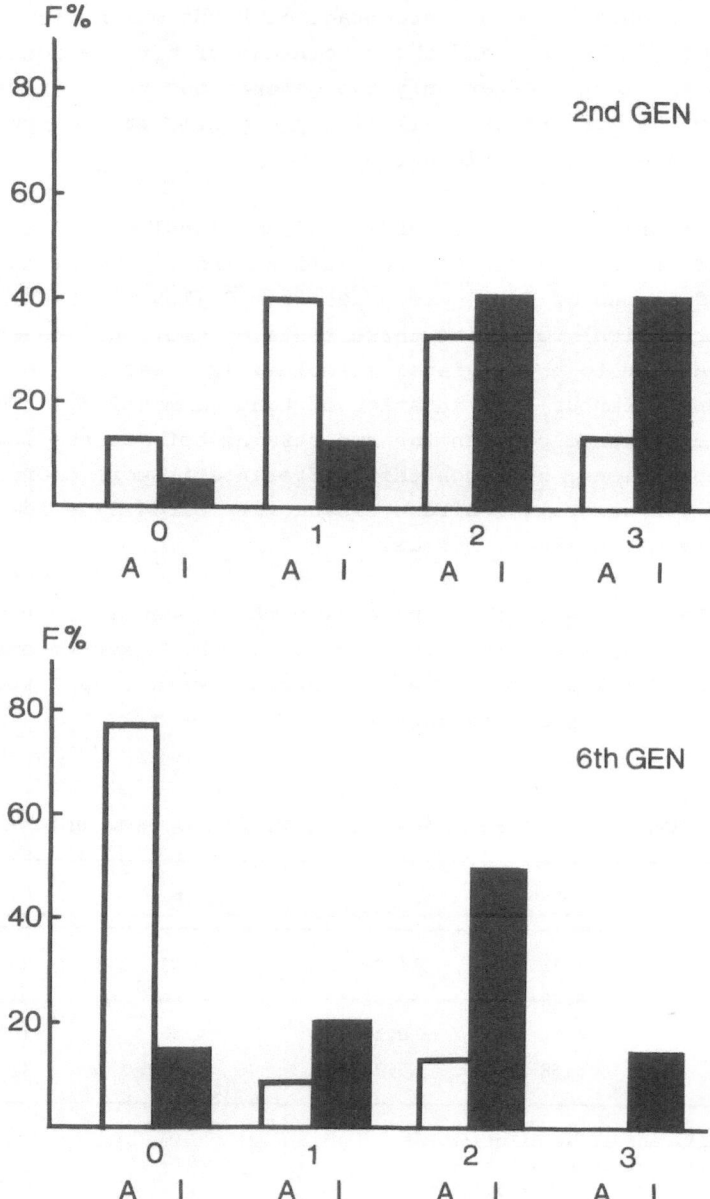

Fig. 2   Distribution of the results of emotional reaction tests carried
out on the 2nd (top) and 6th (bottom) generations. Birds are
classed ·from O (low emotional reaction) to 3 (high emotional reaction)
for the active (white) and inactive (black) lines.

These two measures are directly related to welfare problems because both fear (Duncan and Wood-Gush, 1974) and corticosterone levels (Gross and Siegel, 1973) are increased by intensive husbandry.

Thus, it is possible to select for better adaptation of the bird to modern housing systems. This has also been shown by Brown and Nestor (1973 - 1974) who selected for low plasma corticosterone levels in turkeys.

However, all the above studies were performed in the laboratory.

2. IS IT POSSIBLE FOR A COMMERCIAL BREEDER TO SELECT FOR BEHAVIOUR?

This point, in fact, raises two questions.

1. Is it possible to measure the behaviour of a large number of birds at a low cost?

2. Is it worth doing from welfare and commercial points of view?

Question 1 - Is it possible?

Before selection programmes can be devised we need to record the behaviour of a large number of birds in a relatively short time, but most observations of this sort take a long time. Such observations often involve the reactions of an animal to a conspecific e.g. social behaviour and, because of the inevitable variability of the stimulus conspecific, these observations must be repeated several times. Because of such repetition it often takes as long as an hour to observe one bird (Ortman and Craig, 1968). Obviously this kind of measure cannot be applied to commercial flocks made up of 5 000 - 10 000 birds. Are there, therefore, any short behavioural observations we can make? I believe so, and will give examples.

(a) Open-field

The open area previously described allows very short observations (less than 3 minutes per bird): it is partly (and could be fully) automated. With a fully automated open-field, one persion could measure 500 - 1 000 birds per day (approximately the size of a commercial batch).

(b) Deep feeder

Deep feeders are designed to reduce food spillage but food consumption is often reduced during their first few days in use. This reduced food consumption is clearly due to the fear-eliciting properties of these feeders because it disappears with familiarity and it is more pronounced in nervous birds (Table 2).

The amount of food eaten over the first one or two days may, therefore, be a good way of measuring fear.

TABLE 2

DAILY FOOD CONSUMPTION OF HENS FOR THE FIRST FOUR DAYS WHEN GIVEN DEEP FEEDERS.

	D_1	D_2	D_3	D_4
A	122.5	94.4	140.9	125.2
1	96.5	80.0	108.4*	107.2*

A = active line; 1 = inactive line
* Significant at 5% dilution

(c) Cannibalism

It has been suggested that the presence of blood-stained or mutilated birds is an effective stimulus for cannibalism (Hale and Schein, 1962; Ferguson, 1968) and this implies an attraction towards blood in domestic birds. However, Jones and Black (1979) found a marked avoidance of conspecific blood by young chicks. An experiment is currently in progress to determine whether this avoidance of blood disappears with increasing age and the onset of cannibalism. Anyway, as there is both

olfactory and visual perception of blood it should be possible
to select for avoidance of blood. From this point of view it
is interesting to note how effectively colour preferences were
selected for in quail (Kovach, 1978).

(d) Perching

It has been suggested that perches may improve welfare in
fowl kept in high densities (Hughes and Elson, 1977) or in
certain types of cages. However, this behaviour shows both
marked inter-strain differences and intra-strain variability
(Faure and Jones, unpublished observation) which demonstrates
that perching behaviour is under genetic control and could be
open to selection. It is also easy to measure this behaviour
in a large number of birds.

The above section is conjectural and all these propositions
need further research before they can be applied in practice.
However, they have served to show that it is possible to measure
behaviour.

We must remember that at least two selection programmes
have been carried out on behavioural traits under commercial
conditions; they are, tameness in the fox (Belyaev, 1979) and
the suppression of broodiness in hens.

Question 2 - Is it worth doing?

There are two types of traits which may be submitted to
selection: (a) those which are expressed only in certain
conditions, e.g. perching, and can only occur when perches are
available, and (b) those which are expressed in all breeding
conditions, e.g. fear.

For the first type of behaviour pattern a selection
programme would have to assume knowledge of conditions in 10
years time but we do not have such knowledge. It may, therefore, be better not to select for this kind of behaviour
because when the selection effects finally appear, conditions
may have changed.

It is improbable that any breeding condition could entirely suppress characteristics such as fear, in particular, but also aggression or cannibalism, and thus a selection programme may be worthwhile, but not necessarily from a commercial point of view.

Another problem is that the selection pressure on behavioural traits may not be available for other characteristics and in so far as welfare is not defined in terms of production, we cannot expect that welfare problems will be taken into account in breeding plans. It is clear that broodiness, which is the only behavioural trait consciously submitted to selection in the fowl, has been selected simply because it is closely related to laying. Until now, competition between commercial breeds has focussed only on characteristics such as growth or laying rates. We can only expect a breeder to express interest in welfare problems when other 'commercial' traits reach a plateau, which seems to be the case for laying rate.

CONCLUSIONS

It is possible to select for behavioural traits related to welfare problems and it is also possible to devise behavioural measures which are not time consuming and which may be applied to large numbers of birds.

The problem concerning their application to commercial strains are:

1. A need for further research in this field.
2. The fact that it may be worth selecting for behaviour from a welfare point of view but not necessarily from a commercial one.

In answer to the question posed by the title, the environment certainly needs to be improved, but it is also likely to be worth improving the bird's adaptation at the same time. It is possible, for example, to reduce fear-eliciting stimuli but not to suppress them. Thus, a selection programme to reduce fearfulness would probably produce good results whatever types of breeding conditions prevail in a few years time.

REFERENCES

Belyaev, D.K., 1979. Destabilising selection as a factor in domestication. J. Hered., 70, 301-308.

Brown, K.I. and Nestor, K.E., 1973. Some physiological responses of turkeys selected for high and low adrenal response to cold stress. Poult. Sci., 52, 1948-1955.

Brown, K.I. and Nestor, K.E., 1974. Implication of selection for high and low adrenal response to stress. Poult. Sci., 53, 1297-1306.

Duncan, I.J.H. and Wood-Gush, D.G.M., 1974. The effect of rauwolfia tranquiliser on stereotyped movements in frustrated domestic fowl. Appl. anim. Ethol., 1, 67-76.

Faure, J.M. and Folmer, J.C., 1975. Etude génétique de l'activité précoce en open-field du jeune poussin. Ann. Genet. Sel. anim. 7, 123-132.

Ferguson, W., 1968. Abnormal behaviour in domestic birds. In: M.V. Fox (Ed), Abnormal behaviour in animals, 188-207. Philadelphia: Saunders.

Fox, M.W., 1978. From animal science to animal rights. 1st World Cong. Ethol. Appl. Zool., Madrid, 557-568.

Gross, W.B. and Siegel, P.B., 1973. Effect of social stress and steroids on antibody production. Avian diseases, 17, 807-815.

Hale, E.B. and Schein, M.W., 1962. The behaviour of turkeys. In: S.E. Hafez (Ed), The behaviour of domestic animals, 1st Ed., 531-564. London: Bailliere Tindall and Cox.

Hughes, B.O. and Elson, H.A., 1977. The use of perches by broilers in floor pens. Br. Poult. Sci., 18, 715-722.

Jones, R.B. and Black, A.J., 1979. Behavioural responses of the domestic chick to blood. Behav. Neur. Biol., 27, 319-329.

Kovach, J.K., 1978. Gene effects and gene-environment interaction in the early visual preferences and sexual development of the Japanese quail. 1st World Cong. Ethol. Appl. Zool. Madrid, 351-368.

Ortman, L.L. and Craig, J.V., 1968. Social dominance in chickens modified by genetic selection: Physiological mechanisms. Anim. Behav., 16, 33-37.

Siegel, P.B., 1972. Genetic analysis of male mating behaviour in chickens (*Gallus domesticus*). I. Artificial selection. Anim. Behav., 20, 564-570.

DISCUSSION

J.P. Signoret *(France)*

I would like to propose that the discussion could concentrate on three points. 1. Patterns of behaviour observed in poultry, resulting from a genetic programme but in relation to the environment. 2. The ecological niche - there has been a considerable evolution in recent years compared to the way in which hens were reared for centuries. This evolution has only affected the characteristics which are of economic interest. 3. Variability - there is a very wide variability in the presence or absence, and in the intensity of the patterns. It is possible for genetic selection within this variability but there is also a very wide spontaneous variation.

These seem to me to be the three points which have emerged from the two papers we have had.

M. Zanforlin *(Italy)*

Is not the selection for viable economic characteristics also a selection for particular patterns of behaviour, or particular adaptability of the birds? In young chicks, from one week old to 20 days, I have found great differences in learning ability between different strains. There are also seasonal differences; for example, in the spring they learn much quicker than in the late summer. So, I wonder whether the selection for fast growth is not also a selection for the adaptability of the bird to a particular environment.

J.M. Faure *(France)*

Yes, I think so. The problem is that the hen has been a domestic species for thousands of years but until the last forty years there has been very little change in the hen's environment. For centuries selection for production characteristics was sufficient to keep the hen at a level of adaptability which fitted the environment. However, the last forty years, which

is the era of industrial production, has been too short for
behaviour to evole sufficiently as a correlated character.

R. Moss *(UK)*

I would like to ask a question which seems to me to be
very important. Dr. Duncan has talked about 'learned behaviour'
and 'innate behaviour'. Has anyone done any work on complete
deprivation of any sensation of the bird in order to see what
is necessary to stimulate both the innate and the learned
behaviours? Is there some behaviour when there is no stimulus
at all?

I.J.H. Duncan *(UK)*

I do not know of any work which has been done on complete
deprivation; there have been a lot of experiments on complete
social deprivation - trying to raise birds in isolation - and
that leads to very disturbed behaviour. The work was done
originally to try to separate innate from learned behaviour.
In cockerels it leads to very aggressive behaviour; to their
sexual behaviour becoming very disrupted, and to all their
social interactions becoming very disrupted. However, even an
animal which is kept in social isolation will probably learn
quite a lot from seeing itself so that when it comes to see
another bird there is a distinct possibility that it knows that
the other bird is at least something like itself. I don't know
whether there have been any experiments at all in which birds
have been completely deprived, blinded, deafened, and so on.

J.M. Faure

I have done an experiment with deaf birds - they are
quite normal with regard to growth rate, laying rate and
behaviour.

J. Fris Jensen *(Denmark)*

It was very interesting to see your graph showing the two
strains, low and high activity. Have you calculated the

heritability, the level of activity and perhaps also the genetic correlation to weight gain?

J.M. Faure

The heritability is about 0.3.

B.O. Hughes (UK)

I would like to comment on the question raised by Mr. Moss, that of sensory deprivation. The experiment which most completely meets his requirements is one done by Linda Murphy. She kept individual chicks in biscuit tins which were painted matt black inside, so that there were no reflections, for periods up to six weeks, in complete darkness. She was interested in looking at the effect of this on the fear response to humans thereafter. She compared the chicks kept in the dark with chicks kept under normal conditions. At six weeks the behaviour of the deprived chicks was remarkably normal. They did show exaggerated reactions for a week or so, compared with the normal chicks, but within a week or two after that their response was more or less indistinguishable from the control. This experiment suggested that sensory deprivation had little effect.

P.M. Schenk (The Netherlands)

I think I can give some information on the question raised by Professor Zanforlin, about selecting for some behavioural characters when you are selecting for production characters.

In our department we are interested in genotype/environment interactions. Starting with two different populations of laying hens, namely light Leghorns and medium neavy hybrids, we had two environments as you see in Figure 1. We selected hens for egg production in those two environments for seven generations. We were then able to measure aggressive behaviour within the pedigrees of the four lines in two environments. The important difference between the two

environments was that Environment A was a social environment
and Environment B was a solitary one.

SELECTION ENVIRONMENTS

environment A (variable)	environment B (constant)
hens on deep litter	hens in battery cages
hens in groups of 54 (social)	one hen per cage (solitary)
natural light variable	artificial light 14 h. p. d.
variable humidity	constant humidity (r.h. 60)
variable temperature	constant temperature (10°C)

wh 13_1980_26

Fig. 1.   The two environments, used for the selection for egg production in
white Leghorn and medium heavy laying hens.

We had three groups of 20 animals of each line (Figure 2).
They were kept in the same environment and with the same
autogeny. We started with animals with the same life history
and then at the age of about 5½ months I measured their
aggressive behaviour. We distinguished three activities,
threatening, pecking, chasing and the results are shown in
Figure 3. These measurements were made over a period of six
months. Of course there is a problem; it is very difficult
to define the value of threatening compared with pecking and
chasing because when the animal is showing threatening

BREED	NUMBER	DIVISION
WG	60	3 groups of 20 animals
WB	60	,,
MG	60	,,
MB	60	,,

Fig. 2. Material (animals) used for aggressive behaviour.

code:
WG = White Leghorns selected on the ground
WB =   ,,   ,,   ,,   in cages
MG = Medium Heavy hens, sel. on the ground
MB =   ,,   ,,   ,, , ,, in cages

AGGRESSIVE BEHAVIOUR OF 4 POULTRY BREEDS

	THREAT	PECK	CHASE	TOTAL
WG	3759	714	144	4617
	81.42%	15.46%	3.12%	
WB	2865	463	96	3424
	83.67%	13.52%	2.80%	
MG	1562	482	151	2195
	71.16%	21.96%	6.88%	
MB	1438	424	71	1933
	74.39%	21.93%	3.67%	

Fig. 3. Observed aggressive behaviour over a period of 6 months in 4 breeds (lines) of laying hens.

behaviour it is usually to another bird and so you have to
take into account the interaction of threatening between the
two.

The whole period of six months was divided into seven
periods of three weeks and one period of two weeks (Figure 4).
The white columns are the results of the totals of aggressive
behaviour during those periods of the battery lines and the
blocked columns are the totals of aggressive behaviour of the
other lines. In six out of eight periods the score of the
ground lines is higher than the score of the battery lines.

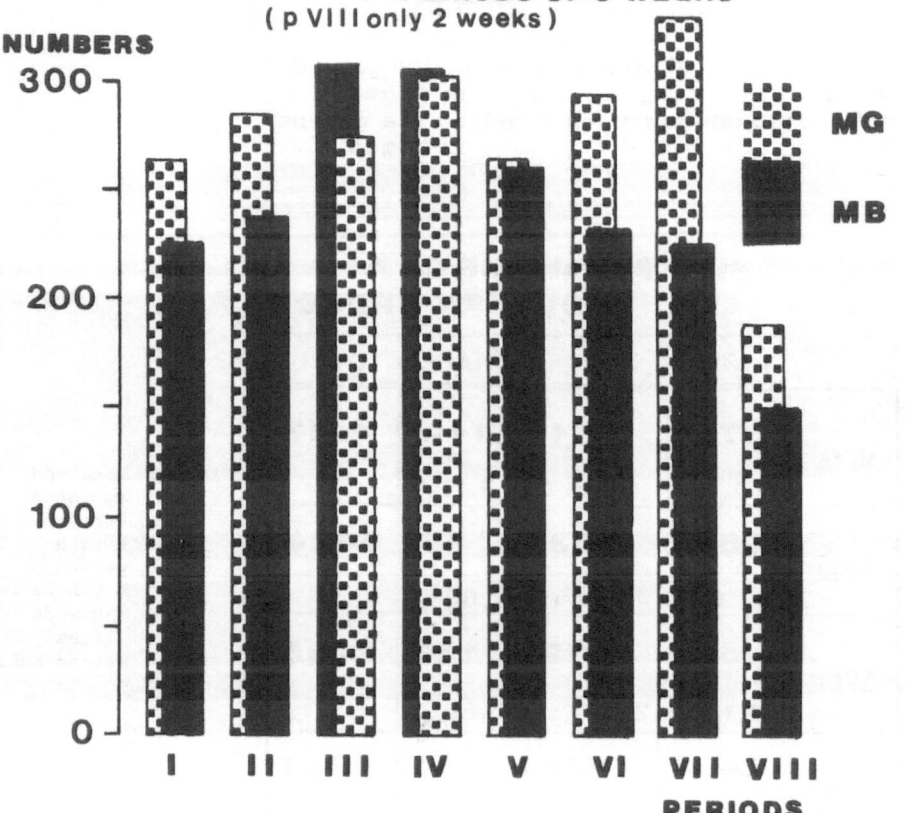

Fig. 4. Aggressive behaviour of medium heavy hens over 8 successive periods

Figure 5 shows the results for the White Leghorns. In this case it is much clearer; the birds from the social environment were much more aggressive than those from the solitary environment.

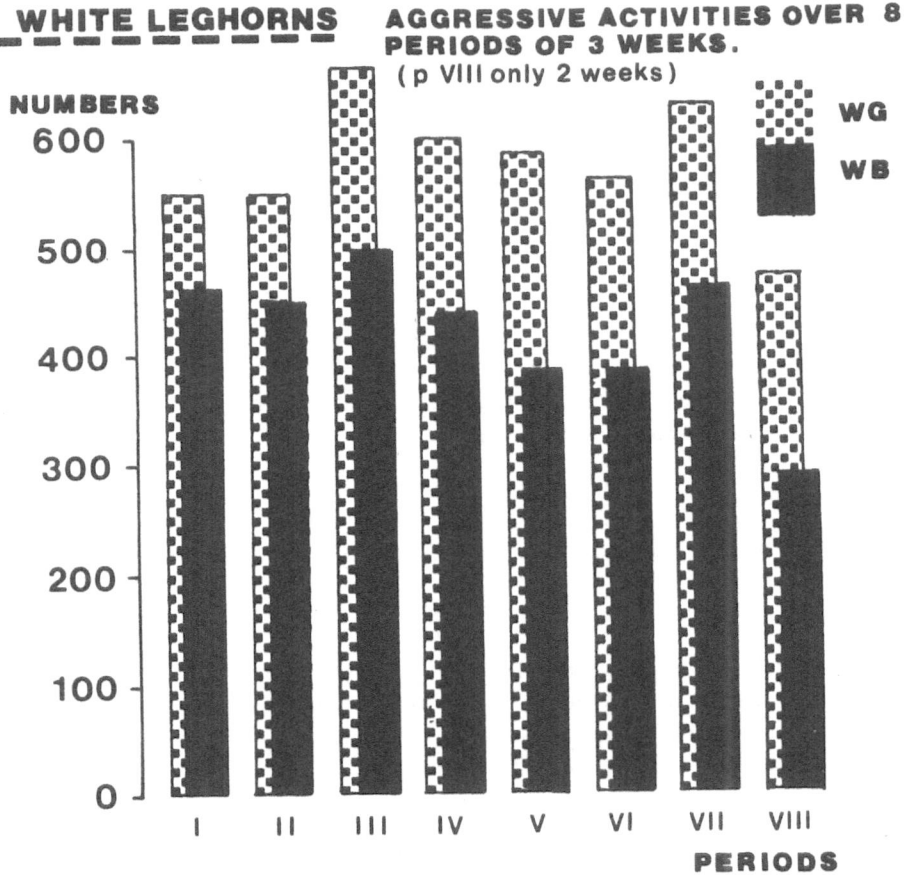

Fig. 5. Aggressive behaviour of White Leghorn hens over 8 successive periods.

I should add that the measurements were made in the groups of 20 animals, not on individual birds. The results show the overall level of aggression in the groups. Observations were made for a period of a quarter of an hour every week during the six months.

M. Zanforlin

What is the precise definition of a solitary environment in this case?

P.M. Schenk

The animals were kept in individual cages; they could perhaps see and hear each other but they had no physical contact.

J. Petersen *(FRG)*

Have you calculated the genetic correlation between aggression and egg production?

P.M. Schenk

No, at this stage I was only interested in the level of aggressive behaviour between the lines.

J. Petersen

Were the interactions significant for aggressiveness?

P.M. Schenk

Yes, I think you must say so. It works in this way: in the social environment there is competition for such things as food, laying nests, and so on. You do not have that in the cages. I think that is the working principle in this case, that you are not selecting directly for production but rather for hens which are high in the rank order which have all the advantages of their rank position.

J. Fris Jensen

May I return to the first paper. As I understand it there are two crucial points. The first one is domestication of the hen; the second is the last 30 or 40 years of intensive selection for productive characters. Can you estimate which of the two is most important?

I.J.H. Duncan

This is an extremely important point. One thing I would
like to point out is that it is reckoned that the domestic
fowl has been domesticated for about 5 000 or 6 000 years but
for the vast majority of that time any selection that there
has been has not been for production characteristics at all.
Generally speaking, the hen has been kept either as a decorative
animal or for its fighting abilities. It is the one domestic
species which has not been bred for tameness. It has been bred
for aggressiveness. That is one thing we must remember all the
time. It is only since about the middle of the eighteenth
century that it has been kept for egg and meat production. It
is only in the last 40 years that intensive selection has been
carried out. However, what the relative importance of these
two things is, I would not like to say. Perhaps Jean Faure
would comment.

J.M. Faure

I think it is very difficult to say. It is very clear
that a lot more has happened in the last 40 years than in all
the time before.

R-M. Wegner (FRG)

I have a question for Dr. Faure. You mentioned experiments
against cannibalism. If you selected against cannibalism, did
you observe a reduction in some production traits?

J.M. Faure

I have not done these experiments myself; I was only
indicating what can be done.

C.M. Hann (UK)

Returning to this question of the relative importance
of the natural and partly artificial selection of 5 000 years
compared with the last 40 years, I think we have to refine our
consideration because it depends on which characteristics we

examine. We have imposed selection criteria or environments
which are so different that in respect of some traits we have
made tremendous changes. The broiler springs to mind where in
the past 20 or 30 years the growth potential has been enormously
shifted from the norm. In other traits, I think we can say that
changes have not been very great. I think we can say that this
is also true with the laying fowl. So if we want to consider
the relative importance of the last 40 years and the previous
many thousands, we have got to look separately at different
aspects of the bird's behaviour and performance.

R. Moss

With regard to the genetic make-up of the bird, there are
two components - the innate behaviour and the learned behaviour.
Both of these, genetic in basis, are modified by the environment.
You then have the adaptive behaviour. I would like to ask both
Dr. Faure and Dr. Duncan, what, within the adaptive behaviour,
are the components from the innate and learned behaviour? How
much is from each?

J.M. Faure

I think the answer may be that heritability is 0.3 which
means there is 0.7 for the environment and 0.3 for the genetics.

W. Goldhorn (CEC)

Dr. Hughes mentioned results of experiments keeping some
hens in isolation and others in a social environment. It
appeared that their behaviour patterns afterwards were almost
the same. Would this not indicate that the most important
factors are innate?

B.O. Hughes

The question that Mr. Moss has posed is a very difficult
one; in a sense it is unanswerable. Dr. Faure answered it by
saying that in his environment there is this ratio of 3 : 7,
inherited : acquired. In the debate on humans regarding the

relative effects of heredity and environment on IQ, it is now fairly well established that one cannot make hard and fast statements. The heritability of a particular ratio is only applicable in a particular environment; if you change the environment you change the heritability. So what you have is two pieces of string; the total length is fixed but the length of the two pieces is variable. There is no way in which you can make a definitive statement. So, in a sense, the question cannot be answered unless you are prepared to specify the environment and then stick to it.

J.A. Hill (UK)

There has been some speculation about selecting birds for various behavioural traits. The conclusion was that from a commercial point of view this probably is not a viable proposition. May I ask the two speakers what they consider to be the suitability of existing breeds to particular environments. I am thinking in terms of one particular behaviour pattern. Ninety five percent of our birds in the UK are kept in cages at present. If we were to change to an alternative system such as litter, what about nesting behaviour and floor laying? Are certain breeds more suited to cages, for example, so that they would not lay in the nests?

I.J.H. Duncan

Of the existing strains, there is evidence that some strains are better suited to cages. One would therefore guess that some strains would be better suited to deep litter. We do not have very much information on what factors are important in nest site selection but there are strain differences in this characteristic. With regard to whether it is worthwhile for commercial companies to select for behavioural characteristics, to take the example of feather pecking and cannibalism, of course it is not worthwhile for commercial companies to select against this characteristic as long as they are allowed to take short term solutions. As long as they are allowed to cut off the beaks of birds, as long as they are allowed to turn down

the lights until the birds are in almost total darkness, why
select? Short term controls are available. Steps could be
taken to make them take this into account in their selection
programmes simply by prohibiting some of these short term
solutions.

P.M. Schenk

I would like to ask Dr. Duncan a question about dustbathing. I think it is very important that we know precisely
what is the function of that specific behaviour. Is it possible
that its function is to remove dead particles of skin from the
bird? We have started a new experiment at Wageningen to try
to establish more information on this.

I.J.H. Duncan

I think your theory is possible. However, I know that
Klaus Vestergaard is going to deal with dustbathing in detail
so I suggest we leave any discussion on that until we hear his
paper.

J.P. Signoret

If there are no more specific questions now we will move
on to the next paper.

SOME CONSIDERATIONS REGARDING OPTIMUM ENVIRONMENTAL CONDITIONS FOR LAYING HENS IN CAGE MANAGEMENT

J. Petersen

In order to obtain eggs for consumption, laying hens today are mainly kept in cages which were developed over the past decades. For the assessment of the various different types of cage developed attention was focused on the laying hen itself and its productive capacity. The hens' laying performance and survival rate, in particular, were used as the main criteria. In this connection, special significance was attached to laying performance, insofar as it is necessary for the various functions of the endocrine system to work together undisturbed for egg formation.

Situations, which place stress on the birds, trigger off hormonal reactions, which affect the control mechanism of egg formation and, as experience has shown, give rise to a drop in performance. If a hen is subjected to environmental stress which it cannot escape, adaptative mechanisms are stimulated which, according to Selye, cause it to adjust to conditions.

If this adjustment fails and more stress is added, weakening the bird's system, this situation can very quickly result in death. For this reason, in the comparison of various types of management, the mortality of a flock of laying hens can be taken into consideration for the assessment of the system.

In modern poultry management, hens and other birds, both in cage and floor management, are kept, as a rule, in a small space. McBright (1970) says on this that, today, poultry lives first and foremost in a social environment, and this social environment is partly determined by physical conditions and the type of management system. In modern management systems new

R. Moss (ed.), The Laying Hen and its Environment, 43-64

criteria apply to hens' living together, which cannot be assessed by the fighting and social behaviour of free-range hens.

In modern housing systems with large animal units attempts are being made to keep social conflicts to a minimum. In cage management, a certain amount can be achieved through the choice of groups. Taking mathematical considerations as a basis, with increasing group size, the increase in the chances of different individuals interacting in the group is disproportionately, large according to the formula $f = \dfrac{n \times (u-1)}{2}$.

Table 1 shows the number of possible interactions between the various hens in a group with 1 - 12 hens per cage, and the disproportionately large increase related to changes in group size is clearly shown.

TABLE 1

RELATION BETWEEN DENSITY AND INTERACTIONS IN GROUPS OF BIRDS

Birds/cage	Number of possible various interactions
1	0
2	1
3	3
4	6
5	10
6	15
7	21
8	28
9	36
10	45
11	55
12	66

If this calculation model is transferred to a constant flock size of 12 birds (Table 2) which is subdivided into different groups, it emerges that the difference between small and large units, as compared with Table 1, may be less but is still clear and pronounced. This simple calculation, however, is only of limited value because in it the important factor 'time' and the chance of repeated interactions in the unit of time, as well as the character of individual birds were not taken into account. However, it reveals one of the causes which helps in the clarification of test results, obtained in the past, on the optimum group size in cage management.

TABLE 2

THE INFLUENCE FROM GROUP SIZE OF THE NUMBER OF INTERACTIONS BETWEEN BIRDS

Flock size	Birds/cage	Number of cages	Possible various interactions
12	2	6	6
12	3	4	12
12	4	3	18
12	5	2.4	24
12	6	2	30
12	12	1	66

Colony cage management was a failure; the larger number of birds in the cage led to poorer performance results. In an experiment by Allen and Perry (1975) it was revealed (Table 3) that the frequency with which cannibalism occurs in cage management is related to the size of the group.

In cages for six the frequency of cannibalism is about double that of cages for three and four.

It should be noted that a cage size of 450 or 650 cm^2 per bird does not affect this feature, as can be seen from Table 3. The studies of Hill and Hunt (1978) (Table 4) also show that as the number of birds per cage increases so too does mortality, egg mass output declines and nervousness and feathering are also adversely affected by increasing group size.

TABLE 3

CANNIBALISM DURING LAY (n = 2 160 birds)

Floor area per bird (cm^2)	Group size	13 - month total	4 - month total
658	3	26	15
658	4	34	20
658	6	62	37
439	3	37	17
439	4	18	11
439	6	51	32

In the current discussions on the cage management of poultry, special significance is attached to the space require-ments of a caged hen. In the past 20 years numerous studies on this subject have been carried out all over the world and, using as a basis the criteria described above, they have reached quite clear conclusions. In 1970 Wegner compiled a very careful and comprehensive bibliography of the results of experiments on the intensive cage management of laying hens. She came to the conclusion that optimum performance with low mortality was achieved with cage floor areas of 400 cm^2 and sometimes even of 350 cm^2 per hen if there was an adequate supply of water and if the length of the sides of the feeding trough amounted to 10 cm per bird. In a large-scale experiment Scholtyssek (1974) tested various types of cage with group sizes of 2 to 6 hens, feeding trough lengths of 6 to 17 cm per bird and stocking rates of 360 cm^2 to 800 cm^2 per bird. An analysis of the different types showed no significant differences in laying performance and mortality in respect of the different features giving rise to the variations. Only a low positive correlation was found between trough length and laying rate (r = 0.2). On the basis of his investigations Scholtyssek came

TABLE 4

THE EFFECT OF DENSITY AND NUMBER OF BIRDS/CAGE ON SOME TRAITS FROM LAYING HENS

Treatments	Mortality (%)	Egg mass per bird (kg)	Feed conversion	Nervousness[c]	Feathering[d]	
					Back	Underside
Density (cm^2/bird)						
310	13.2^s	17.40^r	2.72^s	2.36^r	1.66^r	1.95^r
387	9.8^s	18.03^s	2.56^r	2.04^{rs}	1.88^s	1.98^r
464	5.8^r	17.82^s	2.52^r	1.83^s	2.17^t	2.12^s
Population (no. of birds)						
3	7.1^r	18.06^s	2.52^r	1.63^r	2.09^r	2.15^r
6	9.9^{rs}	17.82^s	2.59^r	1.84^r	1.93^s	2.04^s
12	11.8^s	17.40^r	2.70^s	2.76^s	1.69^t	1.88^t

c = not nervous, 5 = highly nervous;

d = 0 - 33%, 2 = 34 - 66%, and 3 = 67 - 100% feather covered;

r, s, t, = means followed by the same letters not significantly different (P < .01).

to the conclusion that 10 cm could be regarded as a desirable
trough length per hen.

It was not until studies with even less space per hen
that the limits of adaptability were exceeded. On the basis of
their studies of laying hens in cages for 3, 4 and 5 with space
availability of 290, 348 and 406 cm^2 per bird and on the basis
of productivity and mortality Russler and Quisenberry (1970)
concluded that the lower biological limit at which the birds'
performance is affected is 348 cm^2 per bird. The studies of
Hill and Hunt (Table 4) also show that at the very low space
rate of 310 cm^2 an adverse effect on the performance parameters
and behaviour of the birds, measured by nervousness, can be
observed.

As can be seen from the literature, laying hens react to
stresses caused by the management system in their laying
performance and mortality rate.

From the point of view of available space the limits of
maximum physiological stress are relatively low. It is, however,
more important than space measurements for all the hens in one
cage to be able at the same time to feed, sleep lying down
undisturbed at night and stand upright.

Accordingly the space requirements of a hen in a group
cage are, on the one hand, dependent on body size, but also -
as stressed by Duncan in the discussion - on specific activity
of one kind.

Other environmental factors will not be gone into further
here although there would be a great deal worthy of discussion
as regards climatic conditions. The Zentralrerband der
Gelfügelwirtschaft (ZDG) (Central Poultry Industry Association)
has issued a recommendation for German laying hen farmers in
which limits and criteria based on current scientific findings
and relating to the most important environmental factors are
given for the optimum cage management of laying hens. These
recommendations appear in an annex to this paper.

REFERENCES

Allen, J. and Perry, G.C., 1975. Feather Pecking and Cannibalism in a
    Caged Layer Flock. Br. Poult. Sci., 16, 441-451.

Hill. A.T. and Hunt, J.R., 1978. Layer Cage Depth on Nervousness,
    Feathering, Shell Breakage, Performance and Net Egg Returns. Poultry
    Sci. 57, 1204-1216.

Russler, P.L. and Quisenberry, J.H., 1970. Responses of Caged Layers to
    Population Size and Bird Density Stresses. Poultry Sci. 49, 1433.

Scholtyssek, S., 1974. Space Requirements and Performance of Caged Laying
    Hens. Archiv. f. Geflügelk. 38, 27-31.

Wegner, R.M., 1971. Intensive Farming of Poultry and in Particular Cage
    Management. Verlag Eugen Ulmer/Stuttgart.

ZDG recommendations, 1979. Poultry Industry Yearbook. Verlag Eugen Ulmer/
    Stuttgart.

ANNEX

MANAGEMENT CRITERIA FOR THE COMMONLY USED CAGE MANAGEMENT OF LAYING HENS IN CLOSED SYSTEMS

Recommendations of the ZDG

1. CLIMATE OF THE HEN HOUSE

Air rate

In the case of compulsory ventilation, the ventilation equipment must be set in such a way that the air rates for summer according to DIN[*] 18 910 (climate in closed houses) are achieved. The minimum air rates per bird in m^3/h are to be determined depending on the weight or the number of birds, and the target size of the difference in temperature between the air inside and out side ($\Delta t'$) the house (see table). The target size of the temperature difference will be given in the Kelvin scale (k, 1 k = 1°C). (The air rate specifications should be noted when ventilation equipment is installed).

Air temperature

The optimum temperature range in the house lies between $+ 12^{\circ}$ and $+ 22^{\circ}$C. However, production is not reduced if the temperature exceeds or drops below (these limits) for short periods.

Relative humidity

Relative humidity should not fall below 60% or exceed 80% for any great length of time.

Air flow

The speed of air flow near the birds should not exceed 0.3 m/sec if at all possible. Even in summer with higher air temperatures in the house, air flow speeds of over 0.5 m/sec must be avoided.

(* Deutsche Industrie-Norm - German Industrial Standard)

Noxious gases

Concentrations of gas in the air in the house should not exceed the following values:

Ammonia (NH_3): 0.05% by volume (50 ppm)

Carbon dioxide (CO_2): 3.5% by volume (3 500 ppm)

Hydrogen sulphide (H_2S): 0.01% by volume (10 ppm)

2. LIGHTING

Light intensity

Where the birds are kept, the lighting intensity should be at least 10 lux. Even lighting throughout the house or cage layout must be ensured. Filament bulbs or low-pressure fluorescent lamps are recommended as light sources.

Length of lighting periods

The duration of lighting must be adjusted according to the age of the layers. In the last few months of laying it should not exceed 18 hours.

3. FEEDING

A length of at least 10 cm along the side of a trough must be available to each bird. From each cage there must be two nipple drinkers within the reach of each hen. The dimensions of drinking channels correspond, in general, with the length given for feeding troughs.

Feeding and drinking equipment must be checked at least once a day to make sure that it is in working order.

4. CAGE DIMENSIONS

Height

The cage must be at least high enough for the birds to be able to stand upright in all places. In the area towards the back of the cage, the minimum height should not be lower than 35 cm.

Depth

The depth of the cage should be 45 cm (if possible), and in no case less than 40 cm.

Width

Each bird must have available to it a width of at least 10 cm along the feeding place.

Cage floor

Mesh (grating) and wire gauge must be such that the birds can stand on at least three toes of each foot. The angle of slope of the floor should not exceed 12°.

5. NUMBER OF TIERS

So far, birds have not been known to suffer any detrimental consequences from cages being stacked in 1 to 6 tiers in each row.

Observation and inspection of all tiers must be ensured.

6. HEALTH OF THE BIRDS

The general health of the birds must be tested on the basis of laying performance, which should correspond to the age of the birds.

Veterinary inspection and observation should be carried out without delay in the case of externally visible symptoms of illness. Numbers of any losses should be recorded and the cause of death ascertained.

7. HYGIENE AND CLEANLINESS OF THE HOUSE

As a pre-requisite for keeping the stock in good health, impeccably hygienic conditions and corresponding cleanliness must be ensured throughout the house at all times.

DISCUSSION

C.C. Brantas *(The Netherlands)*

Dr. Petersen, you said that egg production is a good parameter for measuring adaptation. You will not be surprised to hear that I do not agree with you. I have two arguments against the use of this production parameter. The first is that one must differentiate between two possibilities. Where you are comparing two animals in the same environment and where you are comparing two farms with the same housing systems, then it is permissible to use production as a parameter for adaptation for welfare. However, if you are comparing two different housing systems, for example, battery cages and deep litter housing, then you are comparing two systems which are both used in agricultural practice and it follows that in both systems production must be good or they would not be commercially viable. Therefore, in that situation it is not logical to use production as a parameter for adaptation for welfare.

My second argument is this: laying hens are forced into egg production by their heredity, by the food they eat and by the lighting system imposed on them. In every way they are stimulated to egg production and they will continue to lay eggs even in conditions which are very bad from the point of view of animal welfare.

We must not make the mistake of saying that because a bird is forced to egg formation by these physiological factors, its environment is necessarily satisfactory from the ethological point of view.

B.O. Hughes *(UK)*

I would like to ask Dr. Brantas how he would interpret the following finding: it was reported recently by Rowland and Harms in Poultry Science that they had a flock of laying hens which had been kept in cages for 15 months. At the end of that time egg production had fallen to about 55%. Half the flock was then translocated from the cages to floor deep litter pens. Contrary to expectations, egg production did not fall, it rose from 55% to 65% within about 14 days. It did not change in the birds which remained in the cages. They interpreted this in terms of the physiological long term stress on the cage housed birds which had been released by movement to the floor, and in this case they took production as a good index of welfare.

C.C. Brantas

It is difficult for me to interpret new data when I am not prepared for it. It would be easy for me to say that I am not a physiologist or an agriculturist, I am an ethologist, and therefore you must ask a physiologist to interpret these data. However, you are talking of a situation in which birds are transferred from one system to another. That is quite different from comparing birds in two different systems at the same time. When illogical arguments are made on the subject of battery cages then I refuse to accept them.

J.P. Signoret *(France)*

At this point in the discussion I think it is difficult to say whether we are for or against battery cages, deep litter systems, or any other system. We have to consider all of them. However, as far as I understand the literature where there has been a change from battery to deep litter, it has not necessarily been followed by an increase in egg production. I believe there is some controversy in this area and there are some results which are opposite to those quoted by Dr. Hughes. Can the physiologists tell us whether it is possible for an animal to adapt to the production it is giving; if not whether an animal having a given possibility for production and prevented from giving his production by environment undergoes some obvious stress?

J.M. Faure *(France)*

If you talk about production in relation to welfare, it is only conjectural. However, you can say that birds are able to adapt to a certain amount of stress in their environment;

that is obvious. If stress goes beyond a certain level then
production will decrease; that decrease in production, of
itself, indicates that stress is present. However, I agree
with Dr. Brantas that by ethological means it is probable that
you can detect lower levels of stress which may be present
although they do not cause a decrease in egg production. This
is important.

M. Zanforlin (Italy)

I think the problem is to find an objective index of
welfare and then to try to correlate it with production.

J. Fris Jensen (Denmark)

May I return to the question of physiology and optimum
environment. I think that temperature is important when we are
trying to investigate different systems. Dr. Petersen mentioned
12⁰ and 25⁰ Celsius. In the two papers by Payne and others
did they give a description of the environment? Was this
research work made on cages or on deep litter because there may
be a difference in temperature requirement between the two
systems?

J. Petersen (FRG)

I do not know about the environment in these two papers.
However, I agree with you that the effect of temperature varies
between the two systems. In cages where there is a high density
of birds they gain heat one from the other. On the other hand
they cannot create heat by sitting on the litter. In warm
climates problems of overheating can occur where there is a high
density of birds in cages.

On the point raised by Dr. Brantas, I agree that egg
production per se is not a criterion for welfare. However, I
think it is the best long-term physiological parameter we have.

K. Vestergaard *(Denmark)*

Obviously it is difficult to make a judgement on the connection between production and welfare. At the present time I think it would be most useful to look at the evidence and see in which cases the two follow each other. We have already seen some cases in which they do. If we look at behaviour we can say that more space is better for the hen, also more feeding space and probably smaller group size. We have seen examples of decreased production when space, and space at the feeding trough were reduced. In these cases fearfulness increased. So let us collect such examples and see in how many cases we have correlation.

M. Zanforlin

I have heard that there is some research being done where the hens can modify the environmental parameters themselves. They can choose temperature, humidity, etc. Does anyone have any information on results of such research?

J.A. Hill *(UK)*

Yes, Stuart Richards at Wye College in London is working on Skinner box techniques where hens can modify their environment as you describe. Obviously only one parameter at a time is being investigated. He has looked at temperature.

M. Zanforlin

Can it be correlated with productivity?

J.A. Hill

I do not know; I don't think he has looked at productivity. I think it is purely a behavioural type investigation.

I.J.H. Duncan *(UK)*

We are doing a little work at the Poultry Research Centre on allowing birds to work in order to get illumination in an operant conditioning situation. The work is in a very early

stage but it looks as if birds will work fairly hard in order
to get illumination, that they will give themselves a pattern
of illumination similar to that which they have experienced
earlier. We have had birds on a 14 hour day length and when
put in a Skinner box they will work in order to turn the light
on for periods of one minute or three minutes and they will
distribute this work throughout the normal 14 hour day. If
they are given a choice of switching lights off or switching
them on, they choose to have more light but, again, they tend
to have their light during the normal daytime. We are now
seeing if they will work in order to switch lights off, if they
will work actively in order to have a dark period. After that
we are going to see if they have any preferences for level of
illumination.

R. Tauson *(Sweden)*

I would like to ask Dr. Petersen if he has any figures on
ammonia concentration in the air, ppm, also if he has the lux
measurement of the light intensity on the different tiers he
showed us. We must be very aware of the fact that depending on
the type of cage there are greater and lesser differences
between the first tier and the lowest tier. Also there are
bulbs today which spread the light more evenly between the
different tiers. In fact, there should not be any difference
in light intensity between the tiers.

J. Petersen

In this investigation there was no measure of lux so I
cannot give figures on variability between the tiers. Neither
was there are information on ammonia. However, I think we can
say that the ammonia level should be less than 50 ppm.

R.M. Wegner *(FRG)*

Did you find that temperature had an effect on cannibalism
in the different tiers? You showed that the upper tier had more
cannibalism because of higher intensity of light. Is it
possible that there was also an influence of temperature because

in many cases there is a higher temperature on the upper tier than the lower ones?

J. Petersen

In this paper there was no information about the temperature between the different tiers. In our own experiment which we have done with different temperatures in houses we did not find an influence. This was with a temperature variation from 32° Celsius at the top, down to 29° Celsius.

C.M. Hann (UK)

I think it is appropriate to mention some work we have done at Gleadthorpe on temperature of layers, where we have adjusted the composition of the diet to match the energy requirements of the birds at different temperatures. When this is done the optimum temperature for a laying fowl, as judged by output of eggs is in the range of 21°C within a band which may extend up to about 25°C. This is borne out when one looks at the kinds of performances which have been possible in semi-tropical countries where by adjusting the diets the birds have become adapted to producing very well at average temperatures much higher than we experience in Europe.

R. Tauson

I would like to return to the experiment with the different tiers. Was it a fact that the reseachers took it for granted that the differences in pecking and other parameters was due to difference in light? What effect did they really study? Or did they only state that they had a difference in pecking on the various tiers. Was it just a tier effect, as such?

J. Petersen

It was a tier effect, not a light effect.

R. Moss *(UK)*

I hope it is in order to cover all this morning's papers
in this discussion. Dr. Duncan stated that some strains do pay
attention to eggs. This was in relation to their possible wish
to incubate eggs. Dr. Brantas has said the hen is forced to
lay by heredity, food and light. The question I am asking is:
is there any evidence in those strains that do pay attention to
eggs, that if the eggs are left to accumulate it will lower
production; that the hen will then begin to go into a stage of
broodiness? Is there a trigger mechanism in the accumulation
of eggs?

J.M. Faure

Of course there is a stop in laying when there is
broodiness but, in fact, if you record the number of eggs laid
by broody and non-broody hens, it is about the same because
just before brooding there is a very big clutch, sometimes about
20 eggs in one clutch. It is the same just after. Very often
the total number of eggs averages out about the same. The
probable reason why broodiness has been selected against by
commercial breeders is that if a bird does not lay for three
weeks it seems that it is eating for nothing but, in fact,
there is not this correlation between broodiness and egg
production.

J.P. Signoret

Are there any more comments on the question Dr. Brantas
raised? It seems very important to me. Dr. Faure showed that
at a certain level of stress there is an obvious fall in egg
production but when production is kept lower than the capacity
of the animal, does this lead to stress? I think this is a
very important problem for the physiologist and the welfarist.
It does not apply only to poultry but to all animal production.

B.O. Hughes

Having asked a difficult question I think I should put my own head on the chopping block! I think that production is an index of welfare but it is affected by a very large number of different factors. It is only if all these other factors are controlled so that they are known to be having no effect on production, that production can be correlated with welfare.

W. Goldhorn *(CEC)*

May I ask Dr. Petersen whether from his review of the literature and from his own work he can say what is the absolute minimum space per bird in respect of physiology and also, perhaps, in respect of welfare? A second question: if we establish that the bird in the present cage system is not in a state of well-being, is it easier to adapt the bird to the cage, or the cage to the bird?

J.M. Faure

In fact it is easier to change the shape of the cage than to select birds. However, we cannot consider animal welfare without taking production into account. We cannot say to producers that they must put all their hens in a field - it would simply mean that we would have to stop eating eggs. We must try to achieve better conditions for the birds without sacrificing high egg production. It may be that selection procedures have a part to play. For example, in the slides shown by Dr. Schenk it was clear that birds selected for high egg production in cages had lower aggressivity.

P.M. Schenk *(The Netherlands)*

I would say that when you are selecting for production in a given social environment you have to take the social characteristics of the animals into account. You may have a lower level of production but you have better balanced birds, they will be less aggressive. In the long term you will probably have better production.

J. Petersen

In reply to Dr. Goldhorn's question on minimum space per bird, I know that it has been stated that the minimum physiological threshold is 348 cm^2, taking into account egg production and viability. However, this figure is open to question, and many factors other than area have to be considered, body weight for example. Various figures have been given, from 320 cm^2 to 400 cm^2. A hen must have enough space to eat, but a smaller hen will require a smaller space. Again, the hen must have enough space to sleep. But it is difficult to give precise figures because the area required inevitably changes according to the body weight of the birds.

I.J.H. Duncan

As a student of animal behaviour I don't think I can accept that smaller hens need less room than larger hens. A hen has a social environment and different strains may well have different requirements. I would doubt very much if it would depend on the size of the hen. It could well be that smaller hens require more room than larger hens.

C.C. Brantas

Dr. Hughes has said that he accepts production as a parameter for welfare in the situation where two housing systems are being compared. I would like to ask him to consider the comparison of two strains within the same house - a good laying and a less good laying strain; for example, a heavy medium strain compared with White Leghorns. Would Dr. Hughes then say that the welfare of the medium heavy strain is lower than that of the White Leghorns? As a second example, one could compare two groups of birds in the same situation but on different feeds. If one group had a feed with just a little bit of calcium in it, then egg production would drop to zero. Would Dr. Hughes then say that the welfare has also dropped to zero? A third example would be a situation where there are two groups of birds kept in the same conditions except with regard to lighting. If one group had one hour per day less light, or had a stepping down system instead of stepping up, there would be a very big difference in production. I cannot suppose that Dr. Hughes would suggest that in any of these three examples there would be any variation in the welfare of the animals.

B.O. Hughes

Of course you are absolutely correct Dr. Brantas. In all those three cases there are other factors which are known to affect egg production, so obviously one could make no assumptions at all with regard to the welfare of the birds because one knows there is an extrinsic effect on egg production. It is very difficult to find situations where you can keep all the circumstances constant as regards their effect on egg production. So, although in principle egg production is a good index of welfare, in practice you have to be very careful to analyse the situation correctly before you can use it as a good index. I would take as an example the shallow cage where we found a consistent increase in egg production with the change of cage shape. As the trough got wider, egg production went up. Going up from 10 cm to 15 cm per bird, the increase in egg production was between 2 and 4%. We did not know whether it was due to an improvement in welfare because we had altered the social structure of the group, or whether it was an improvement in food intake, the conversion ratio affecting egg production directly. We suspect that both factors played a part but it is very hard to be certain.

G. Martin (FRG)

In the experiment you describe, Dr. Hughes, I think you were only improving the animals' well-being marginally with the change in cage shape and the wider trough.

The other point I would like to make is in connection with Dr. Petersen's paper. We should not only be discussing the space available to each animal, 320 or 400 cm^2, but also what else the animal needs. It is not just a question of quantity but also of quality.

J.P. Signoret

I believe Dr. Faure's paper is relevant to this point.

J.M. Faure

We have certainly established that there are strain differences in birds' requirements to perch, for example. Also, it is quite easy to find out what the birds prefer to do but that is not necessarily what is best for them.

J. Fris Jensen

I am a bit puzzled at Dr. Martin's distinction between quality and quantity because when you are saying that you want this kind of thing for a bird you also want to know how much of it. So I do not think that quality and quantity can really be divided. But, of course, if anyone has any suggestions for a new criterion which may be very important for the bird situation then now is the time to put it forward so that we can mention it as a qualitative criterion and then try to measure and make it quantitative.

K. Vestergaard

Dr. Petersen, you said that there was no decrease in egg production as a result of allowing the birds 400 cm^2 per head. As far as I remember, Dr. Hughes did a survey on these things - the effect of flock size, density and so on - and there was an increase up to the highest level studied which was 580 cm^2 per bird. So there is some increase above 400 cm^2.

There is another point: you were talking about group size and you said that if there was only one bird there would be no interactions. But, in fact, it has been established that birds in neighbouring cages have social dominance relationships with each other. There are also records of pecking from separate cages. So there are some interactions.

J. Petersen

Many investigations have been done on the space required for hens and the influence on production. There was a very complex investigation carried out by Professor Wegner which showed that 400 cm^2 did not influence laying intensity. If you take the overall picture of the results in the literature I think the figure of 400 cm^2 is a fair conclusion.

J.P. Signoret

I am afraid we must stop the discussion now.

CORTICOSTEROIDS IN LAYING HENS

G. Beuving

In the literature adrenal hormones have often been related to stress and sometimes equalised as stress hormones.

In the well-known stress concept of Selye catecholamines represent one aspect of the acute phase of the alarm reaction to stressors.

The hypothalamus-hypophysis-adrenocortical axes are at least equally important for homeostasis (Selye, 1973).

In the original concept adrenal enlargement was one of the indices of stress, together with shrinkage of lymphatic structures and ulcerisation of stomach and upper gut.

Because stress is a largely imprecise term and means different things to different people, Brown (1967) supposed that perhaps we can live with it, if we define it as the generic term for those diverse immensely dissimilar stimuli, which have in common the experimentally proved property of stimulating ACTH secretion and in consequence corticosterone secretion. He used plasma corticosterone concentrations in turkeys as a parameter to select a 'low stress' and a 'high stress' line. Also Freeman (1978) says that the key to the response to stressors are the adrenal glands: no matter what the stressor, these glands are stimulated and thereby regulate the response to the stressor. One can wonder if this means that each response of the adrenal glands means that a stressor is implied. However, experimental proof of this proposition is not available. On the other hand, very few reliable data about adrenal hormones were present, especially from birds, neither from glucocorticosteroids nor from cathecholamines. Therefore, some years ago we started a research programme to study

R. Moss (ed.), The Laying Hen and its Environment, 65-84

plasma corticosterone in laying hens and the influence of
several stimuli on this hormone.

Until recently, very few data about corticosteroids in
plasma have been available. Fluorometric methods were
generally used and gave very high and totally different values.
Competitive protein binding and radioimmunoassay techniques
are now available. It should be stressed, however, that any
technique that is validated for measurement of adrenal steroids
in mammals will require additional validation before application
to the quantification of the same steroids in avian blood
(Frankel, 1970).

Application of the method of competitive protein binding
as described by Murphy (1967) for mammals to the plasma of
laying birds without purification of the assay samples gives
erratic results in our hands. Nir et al. (1975) found values
of about 100 ng corticosterone/ml plasma in 21 day old chicks
with the method of Murphy. Buckland and Blagrave (1973) however,
found values of 5 ng/ml in 39 day old chicks also with this
method, but after a better extraction of corticosterone (in
birds there is only one corticosteroid: corticosterone).

Culbert and Wells (1975) found values of 7 - 20 ng/ml plasma
in laying hens, with fluorometric methods, after purification
by column chromatography. Weiss and Brand (1974) found values
of 2.2 ng/ml in plasma of broilers and maximal values of 2.7
ng/ml in plasma of laying hens with competitive protein binding
but without chromatographic purifications. Viewed in this
light the remark of Frankel seems to be very true that any
technique used in mammals will require additional validation
before application to avian blood. Therefore we tested each
change of the technique in two ways:

1. independence of the amount of plasma, used for the
    assay

2. recovery of endogenous corticosterone.

With the method of competitive protein binding we found it necessary to purify plasma samples by column chromatography with Sephadex LH-20. We also used the same purification with radioimmunoassay but this purification step is then less necessary. The RIA technique is more accurate especially in the range of 0 - 3 ng/ml. It also needs less blood volume for the assays.

After validation of the method another important problem was how to get blood samples without disturbing the birds. For that we used cannulation of the wing artery with a polythene cannula of about 150 cm. The cannula is led outside the cage by a pulley, closed by a clamp and filled with a heparin solution (Figure 1). The first part of the cannula is protected by a second tube around the cannula. This outer tube is painted blue to protect the cannula against pecking, especially when blood samples were taken.

In this way, blood samples can be taken during day and night. The presence of the cannula did not seem to disturb the birds at all and they did not react when blood samples were taken. Each day 1 ml heparin solution (500 U) is injected via the cannula to prevent blood clotting.

Despite these injections the cannulas cannot be used for longer than a week and are not suited for experiments that need more time.

In nearly all experiments, the first blood samples were taken at least 36 hours after cannulation had taken place. In all cases laying hens at the age of 25 - 40 weeks were used (White Leghorn).

To see how the concentration of corticosterone changes during a day, blood samples were taken during 24 hours in about 10 laying hens, individually housed in cages. In a first experiment the method of competitive protein binding was used for the assay (Figure 2). An increase in corticosterone could

be shown at the end of the night. During the day there was a
general decrease and the contents were minimal at the end of
the day. The same pattern was also shown in pigeons (Boissin
and Assenmacher, 1970) and in mammals. In nocturnal animals a
reversed pattern could be shown. The daily rhythm was only
found as an average of several laying hens and was not clear in
the individual values during a day. In three cases an egg was
laid during blood sampling. These samples contained higher
values of corticosterone. When they were omitted, the dotted
line was obtained. This was an indication that during egg
laying corticosterone contents were increased. In a second
experiment the daily rhythm was determined again, but now
corticosterone was assayed with radioimmunoassay.

Fig. 1.

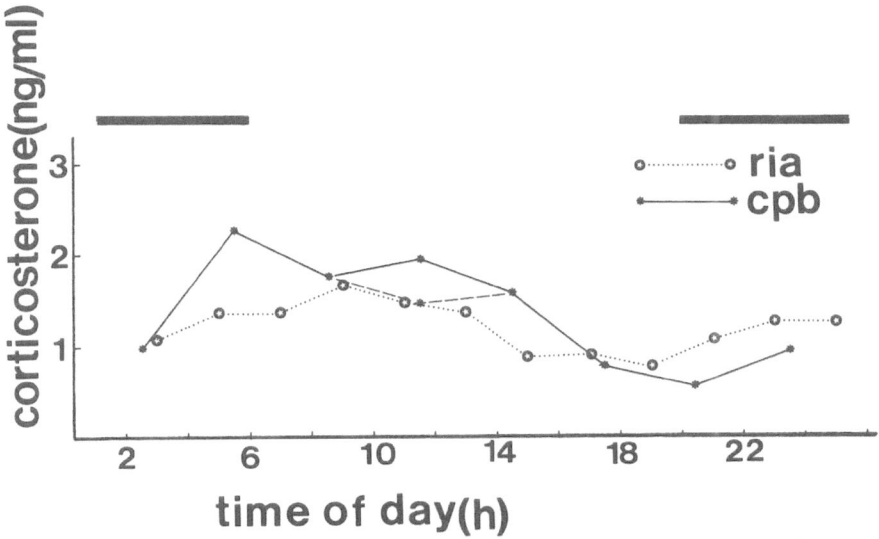

Fig. 2.

However, in this experiment no blood samples were taken
from hens two or three hours before and one hour after egg
laying. No important differences between the curves can be
shown, only the difference between maximum and minimum values
is less distinct in the RIA curve.

The indication that the corticosterone content should be
increased during egg laying was examined in a further experiment.
Laying hens were singly housed in cages. Blood samples were
taken at 20 min intervals starting about one hour before the
estimated time of egg laying. This scheme was sustained, also
when an egg was laid. Thereafter, the time of blood sampling
was reckoned from the time of oviposition. The respective
values were classified in 20 min groups (Figure 3). An increase
starting at least 1 hour before egg laying could be shown.
However, in a cage without a nest a hen cannot perform her
normal egg laying behaviour. For 1 - 2 hours before egg laying
she occasionally shows a state of agitation with escape movements and restlessness. To check if this disturbance of laying

behaviour may cause the increase of corticosterone, cannulated
hens were maintained individually in a cage of double size in
which a wooden nest with buckwheat shells was placed. Under
these conditions the birds show their 'normal' egg laying
behaviour before egg laying and no signs of agitation, escape
movements or restlessness could be observed. The hen sits on
the nest for about an hour. In this case too, we used cannulated
birds. The concentration of corticosterone however, showed the
same increase as during egg laying without a cage (Figure 3).
These results suggest that the increase of corticosterone
before egg laying is not caused by the disturbance of normal
egg laying behaviour.

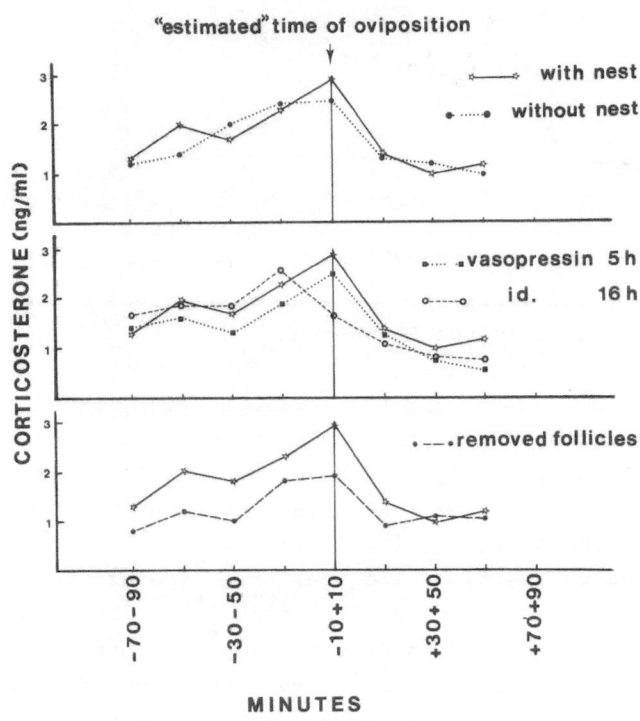

Fig. 3.

Eggs can be laid prematurely by the injection of vaso-
pressin. The egg is laid a few minutes after intra-arterial
injection. The hen normally performs her egg laying behaviour
at the estimated time of egg laying. Again blood samples were
taken at 20 min intervals around the estimated time of egg
laying. The time that the hen leaves the nest is taken as the
'normal' time of egg laying. As can be seen (Figure 3), the
increase in corticosterone does not change when the egg is laid
prematurely 5 h or 16 h before egg laying.

In conclusion, we can say that the whole increase, or at
least the greater part of the increase, in corticosterone during
egg laying does not come from external circumstances, but is a
part of the control of egg laying.

Next we examined the possible influence of ovulation on
corticosterone concentrations. Sometimes we had an indication
that after the increase during oviposition, there was a second
small increase. To prevent missing this little peak, we took
blood samples directly after egg laying, each 10 min instead of
20 min. This experiment was done in two groups (Figure 4).

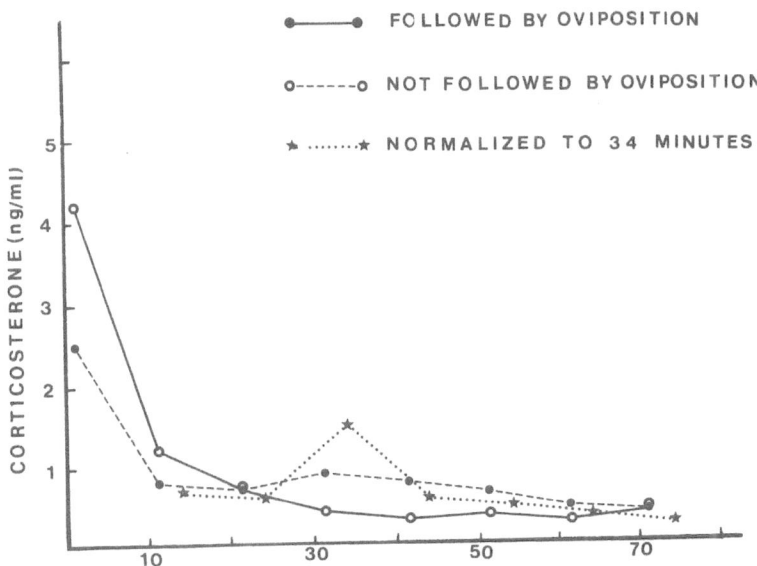

Fig. 4.

1. Birds that have laid the last egg of a clutch, so that ovulation did not happen in the hours directly after egg laying (control line).

2. Birds that had laid an egg in the beginning or in the middle of a clutch.

In both cases the birds were killed after blood sampling to check if ovulation had taken place or not. Combination of values from the second group gave a very small increase. However, one distinct peak value was present in nearly all the series of blood samples from each individual hen. Because the peak values were found at various times after oviposition, only a slight increase was present in the average values. The average time of the combined peak values was 34 min after oviposition. This peak value is probably due to ovulation and agrees very well with 15 - 75 min after oviposition, as indicated by Sturkie and Mueller (1976). When all peak values, due to ovulation, were normalised to 34 min a distinct peak could be shown. In the control line no distinct peak values in the individual hen could be found.

To examine the level of corticosterone during a short period we took blood samples at 2 min intervals for half an hour.

A. In the afternoon, several hours after egg laying.

B. Just before egg laying, when the levels were higher.

As we can see from Figure 5, the concentrations did fluctuate very clearly with a certain regularity. The interval time was about 10 - 12 min. In the period before egg laying high peak values can be shown.

In the first part I described the influence of daily recurring events (daily rhythm, oviposition, ovulation, short time fluctuations).

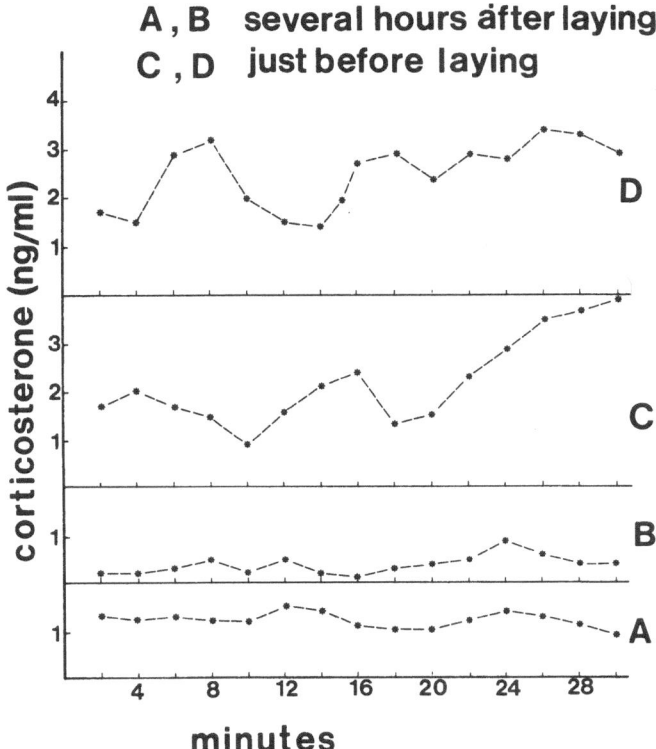

A , B   several hours after laying
C , D   just before laying

Fig. 5.

The next step was to do some research on the influence
of external stimuli on corticosterone concentration.

First we investigated the influence of handling, or
rather immobilisation by hand (Figure 6). The birds were taken
at t=o from their cages and fixed on a table by hand. Blood
samples were taken each ½ minute. There is a strong increase
for 7½ min. A part of this increase, especially the second
part of the curve must be ascribed to a haemorrhagic shock.
But irrespective of this effect it can be seen that there is
an influence of handling, probably within one minute. The
value of 45 sec differs significantly from the t=o value. This

means that it will be very difficult to measure corticosterone
in blood samples, taken by hand, without any influence of
handling.

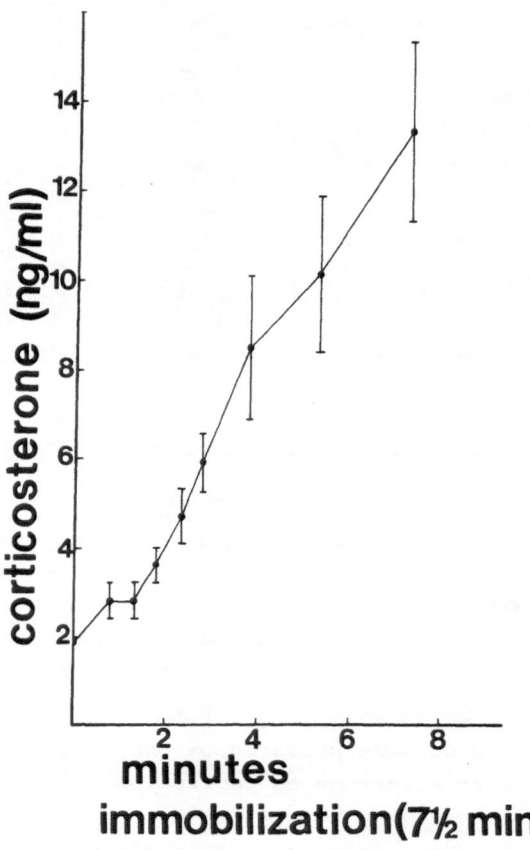

Fig. 6.

When the birds were fixed on the table for a longer
period (1 h), a fast increase during 5 min is followed by a
gradual increase during that hour of immobilisation (Figure 7).

In another experiment we looked at the influence of
repeated handling (for 5 min) 4 times a day for 5 days (Table 1).
Adaptation to the influence of repeated handling was not

TABLE 1

THE INFLUENCE OF REPEATED HANDLING ON 5 SUCCESSIVE DAYS IN NINE CANNULATED LAYING HENS[A]

	First day			Second day		Third day		Fourth day		Fifth day	
Day time	08.00	08.35	12.35	08.35	12.35	08.35	12.35	08.35	12.35	08.35	12.35
Arithmetic mean of corticosterone concentration (ng/ml)	2.3	6.1	6.5	4.0	4.3	4.2	4.8	4.3	4.8	4.0	5.1
± SE of arithmetic mean	0.9	1.2	1.5	0.9	0.8	0.8	0.8	0.7	1.2	0.7	1.2
Retransformed values of the geometric mean	1.9	5.3	3.6	3.0	3.6	3.5	4.1	4.0	2.9	3.4	3.9

[A] The hens were immobilised by hand for 5 min at 08.30, 10.30, 12.30 and 14.30 h. Blood samples were taken twice a day at 08.35 and 12.35 h. On the first day a reference value was determined at 08.00 h.

apparent. The values remained significantly higher compared with the control values, even when the birds had been handled 20 times.

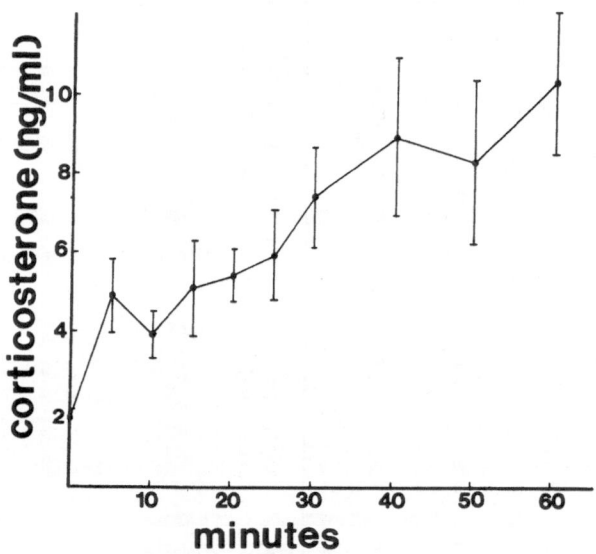

immobilization during 1 hour

Fig. 7.

Immobilisation for a much longer time by crating (for 7 hours) also caused a strong increase. The level is quite constant during the first 4 hours (Figure 8). Thereafter the level tends to increase still more. This may also be influenced by deprivation of food and water.

Increasing the temperature from 20° to 37° during one week increased the corticosterone level. The hormone levels were elevated on the first day, but gradually fell to a lower level during the experimental period (Table 2).

TABLE 2

THE INFLUENCE OF HEAT STRESS (37°) ON 10 CHICKENS FOR 7 DAYS[A]

	First day						Second day		Third day		Fourth day	Fifth day
Day time	08.00	09.15	10.00	11.30	13.15	15.30	09.00	15.30	09.00	15.30	09.00	09.00
Arithmetic mean of corticosterone concentration (ng/ml)	3.1	5.3	4.9	5.0	3.6	2.6	4.0	3.3	4.4	3.8	3.3	3.2
± SE of arithmetic mean	0.63	1.00	0.96	0.80	0.68	0.64	0.58	0.55	0.79	0.41	0.55	0.65
Retransformed values of the geometric mean	2.6	3.5	3.3	3.8	2.0	1.8	3.7	2.5	3.5	3.5	2.8	2.8

[A] Heating started on the first day at 08.30 h after the first samples had been taken.

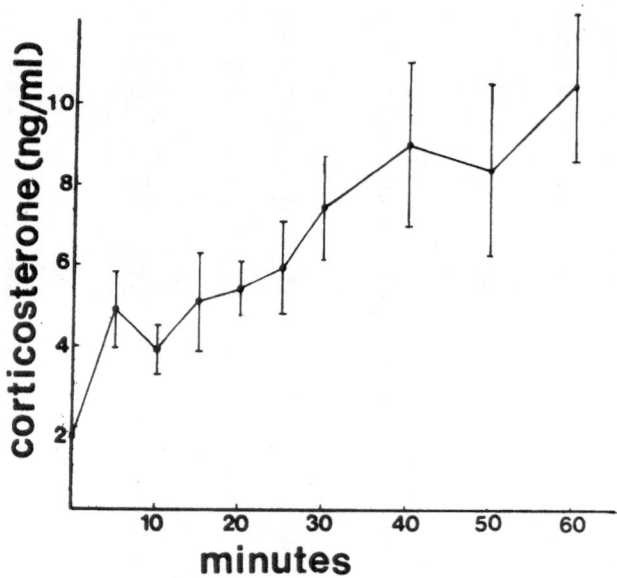

Fig. 8.

immobilization during 1 hour

Also the deprivation of water (2½ days) and food (5 days) had a distinct effect on the hormone level. The combined values of the second and the third day compared with the combined values of the fifth and the sixth day differed significantly (P < 0.01) (Table 3). No effect of starvation could be shown because the level returned to normal after water was given on the fourth day. At the end of the sixth day all hens had stopped laying.

To compare all these values with the maximal production that is possible in laying hens, we made a comparison between the amount of porcine ACTH, injected via the cannula and the concentration of corticosterone. A maximal concentration of about 30 ng/ml plasma could be shown (Figure 9). This maximal concentration was reached at about 0.5 U ACTH/kg in half an hour. It can be suggested from these curves that an increase due to handling once disappeared within a quarter of an hour.

TABLE 3

THE INFLUENCE OF WITHHOLDING FOOD AND WATER[A]

	First day	Second day		Third day		Fourth day		Fifth day		Sixth day	
Day time	07.45	07.45	15.15	07.45	15.15	07.45	15.15	07.45	15.15	07.45	15.15
Arithmetic mean of corticosterone concentration (ng/ml)	3.1	3.8	4.5	4.4	3.2	5.5	3.1	2.7	2.6	2.8	2.3
± SE of arithmetic mean	0.36	0.49	0.52	0.55	0.72	1.04	0.64	0.74	0.44	0.68	0.40
Retransformed values of the geometric mean	2.9	3.6	4.3	4.1	2.4	4.7	2.6	1.8	2.3	2.1	2.1

[A] Water was withheld for 2 1/2 days and food was withheld for 5 days (n = 10). Water and food were removed on the first day at 20.00 h. Water was given again on the fourth day at 08.00 h.

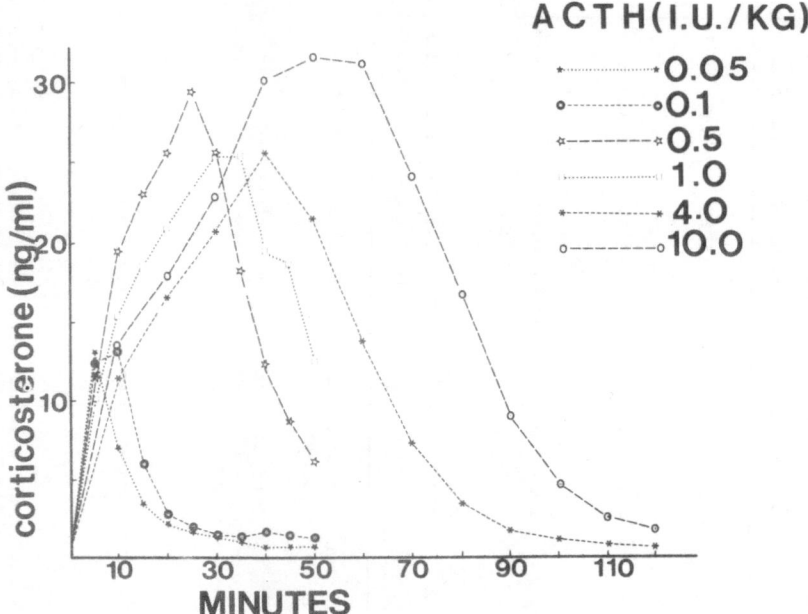

Fig. 9.

Finally we compared the maximum values of each experiment. This is, of course, not an accurate comparison, but it gives an impression of the order of the increase in each experiment. It shows that natural conditions (heat, deprivation of food and water) caused smaller increases than the 'unnatural' effects of handling, crating immobilisation for a longer time etc.

No results can be given from birds housed on the floor, because we were not able to take blood samples without handling the birds. However, the daily rhythm experiment makes it improbable that in the cage birds (1 000 cm^2/bird) the corticosterone content was increased by stress because:

1. the daily rhythm was repeatable

2. very low values were found, especially in the evening

3. there was agreement with daily rhythms in mammals.-

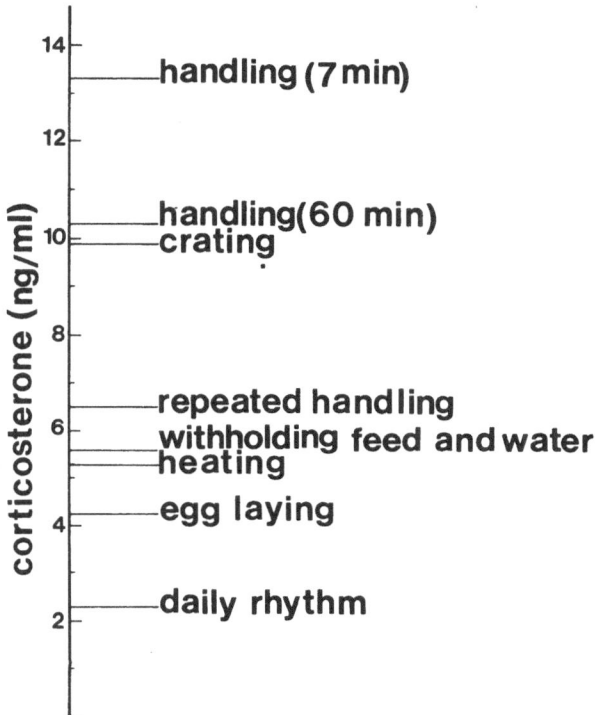

Fig. 10.

The results presented here, do not provide answers to questions about stress in cages with more dense populations.

We hope, however, that these results, together with many other data can give us more insight into the relationship between adrenal hormones and stress.

REFERENCES

Boissin, J. and Assenmacher, I., 1970. J. Interdiscipl. Cycle Res. 1,
251-265.

Brown, K.I., 1967. In 'Environmental Control in Poultry Production'.
(Ed. T.C. Carter), pp. 101-113. Oliver and Boyd, Edinburgh.

Buckland, R.B. and Blagrave, K., 1973. Poult. Sci. 52, 1215-1217.

Culbert, J. and Wells, J.W., 1975. J. Endocr. 65, 363-376.

Frankel, A.I., 1970. Poult. Sci. 49, 869-921.

Freeman, B.M., 1978. In 'First Danish seminar on poultry welfare in
egg-laying cages'. Ed. L. Yding Sørensen, Herlev.

Nir, I., Yam, D. and Perek, M., 1975. Poult Sci. 54, 2101-2110.

Selye, H., 1973. Am. Sci. 61, 692-699.

Weiss, J. and Brand, J.H., 1975. Zentbl. Vet. Med. A21, 225-242.

DISCUSSION

B.O. Hughes *(UK)*

Dr. Beuving, I found your results very interesting; they fit in very nicely with what one might expect, particulary in the case of handling. With regard to egg laying, would it be reasonable to assume the results indicate that egg laying, itself, is a stressor, or is there some physiological reason why corticosterone should rise during the run up towards laying?

C. Beuving *(The Netherlands)*

My interpretation is that you cannot say that egg laying is a stressing factor because there was also an increase of corticosterone when eggs were not being laid.

B.O. Hughes

This fits in beautifully with Wood-Gush's findings, which I am sure you are aware of, where he removed the follicle and the bird went through all the normal behaviour at the time when it would have laid the egg, including searching out the nest, even though the egg had been laid many hours previously. So there would appear to be correlation between the behaviour and the corticosterone but these can both be physiological, you say, and in this case stress is not necessarily a factor?

C. Beuving

Yes, that's right.

B.O. Hughes

Of course, Wood-Gush thinks that the behaviour is produced by a correlation between oestrogen and progesterone. Would it be possible to argue in your case that you think these hormones are involved with the release of corticosterone?

C. Beuving

I can't answer that question; I don't know what is the relationship between these steroid hormones and the corticosterone.

J. Fris Jensen *(Denmark)*

I would like to thank Dr. Beuving for his paper and his very interesting results.

SUMMARY AND DISCUSSION

J.P. Signoret *(France)*

We must now attempt to summarise the first session and
I would suggest, if you all agree, that we should do this by
taking some key words and putting them up on the board as
headings which we can use to formulate our summary.

R. Moss *(UK)*

Chairman, before you do that may I put forward four
questions which seem to me to be important and which, in the
main, have to be directed at Dr. Duncan.

With regard to feeding behaviour, in the two cases that
he has spoken about, the appetitive and the consummatory
behaviour, in the first what are the trigger mechanisms for
locomotion and scratching, and in the second, is Dr. Duncan
saying that it is essential to retain the upper mandible intact
to be able to undertake the consummatory behaviour satisfactorily?

The second question relates to the components of feeding
behaviour. Dr. Duncan spoke of the difference between strains.
I wonder if he can explain how much difference there is, and in
which of the components it occurs.

Thirdly, he talks of adaptive behaviour, and innate and
learned. Which is more important in the make-up of frustrative
behaviour and can he distinguish between what is adaptive and
what is frustrative?

Finally, may I ask for a definition of 'social space'?

I.J.H. Duncan *(UK)*

With regard to your first question, the trigger mechanisms
for locomotion and scratching, there does not seem to be any
trigger mechanism for scratching. It seems to be linked in some
way to pecking; if you get pecking there is a tendency for the

R. Moss (ed.), The Laying Hen and its Environment, 85-97
Copyright ⓒ1980 ECSC, EEC, EAEC, Brussels-Luxembourg. All rights reserved.

bird to scratch as well. However, I have the impression that if the bird is slightly frustrated, if it is not getting food quite as fast as its level of hunger requires, it tends to show scratching. I'm afraid that is not a very satisfactory answer.

M. Zanforlin *(Italy)*

I would confirm that - I have seen it frequently when I have given reinforcement to the animal that is pecking to obtain food. If I withhold the reinforcement it does start to scratch.

I.J.H. Duncan

Yes, we have noticed the same in Skinner boxes where birds are having to perform a pecking response in order to obtain a reward of food. If you make them work hard, if they have to peck often then scratching begins to interrupt the sequence of pecking.

Removal of the upper mandible: we have found that removal of one third to one half of the upper mandible will disrupt feeding behaviour for about three to four weeks. Eventually the bird will learn how to compensate for this but it depends very much on how the food is presented. If the bird is being fed a pelleted diet on a hard flat surface, it will never learn how to pick the food up. It is very difficult of course because of the discrepancy in the length of the bill, the fact that the lower mandible will hit the solid substrate first, it will never be able to grasp a pellet. However, if you put the bird onto a mash diet, where it can actually dip its beak into the diet, it will learn how to overcome this handicap. Food consumption is depressed for about three or four weeks as well but eventually it will go back to its pre-operative level - on mash.

Appetitive and consummatory behaviour: there are differences. To take a very exaggerated example, broiler chickens and the light hybrid laying chicken, the broiler chickens are much more efficient feeders. They show less of the appetitive elements and more of the consummatory elements. The consummatory

elements they show are much more efficient, they take much
bigger mouthfuls of food. They spend a much shorter time in
feeding behaviour whereas the light hybrid laying birds spend
a lot of time pecking and exploring the food and are much less
efficient. That is a big difference between broilers and
layers; there are smaller differences between different strains
of laying birds.

In your last question you have divided behaviour into
innate and learned. I was very careful to say that there is
a whole range of behaviour with, at one end of the spectrum,
behaviour which seems to be buffered from the environment, to
the other end of the spectrum, behaviour which is very open to
environmental influence - although, of course, all behaviour
has a genetic component. If you try to divide behaviour simply
into innate and learned it is a false division which will lead
to wrong conclusions.

R. Moss

So within that range, not dividing it at all, when you
do get adaptation, is there a border line between adaptive and
frustrative behaviour, because you have provoked frustrative
behaviour and yet you speak of adaptive behaviour within the
total range between learned and innate?

I.J.H. Duncan

I have some unpublished evidence that some of the responses
that birds make to mildly frustrating situations may be adaptive.
One of the responses that birds give to mild frustration is an
increase in preening behaviour, displacement preening which is
slightly different from normal preening. We have some evidence
that when the bird performs this behaviour it seems to calm
down. We are now trying to quantify this, looking at its
physiology and short term responses such as heart rate. I do
not have any results yet. If hens are more severely frustrated
they show stereotyped movements; it is possible that these may
be adaptive as well, I am not sure. We do not have any evidence
at the present time but we hope to find some.

M. Zanforlin

I think frustration always precedes adaptation by learning.
Whether the bird is exhibiting innate or learned behaviour, if
a frustrating situation occurs, either the bird will learn to
adapt in which case the frustration will vanish, or it will be
unable to adapt in which case the frustration will get worse.

J.P. Signoret

This last point has been discussed also by Dr. Petersen,
the problem of social space and minimum surface per animal.
Are there any further comments on this.

B.O. Hughes *(UK)*

Social space is difficult to define. It is based on the
idea that McBride first put forward that birds tend to adopt
certain configurations with respect to each other's bodies.
For example, if two birds are facing head to head this might
be more likely to lead to an encounter than if they are facing
head to side. By taking photographs of birds in pens from
above, he found that the head to head relationship was less
common than you would expect to find on a random basis, suggesting that birds avoid this kind of encounter. If you translate
these findings to the cage situation it indicates that the birds
don't just need enough space to avoid each other's bodies, they
may need to be able to orient themselves to be able to avoid
this kind of confrontation which may require more space. The
classical example of this going wrong is in flat deck feeding
where you have central troughs running down between two banks
of cages and birds are feeding facing each other, particularly
where you have restricted feeding where the food comes down and
the birds have to feed while the food is available. In this
situation there is a great deal of aggression against birds in
the opposite cages. This appears to support the findings of
McBride, although there is room for the birds to feed social
space demands will require that the cages are sufficiently
far apart so that this response does not occur.

I.J.H. Duncan

McBride's theory seems to me to be a good one. It is a pity that no one has set out to make measurements and try to put linear measurement on it.

B.O. Hughes

Professor E.M. Banks and I did some work on this last summer, again taking cine photographs from above. He is analysing his results at the moment. They are very interesting and tend to support McBride. I think what we need now is more work in cages rather than in pens.

I.J.H. Duncan

I think we need more basic work to say what is the distance involved during feeding, walking about, or during dust bathing. Obviously, it is much less during dust bathing; birds come into physical contact with each other there. It must depend on the activity the bird is performing. If we have the basic measurements then perhaps we can say more about designing environments to suit these measurements.

J.P. Signoret

I was impressed by the fact that the reports from Dr. Duncan and Dr. Faure showed that it is possible to describe an ethogram which shows a wide difference between a wild fowl and the modern laying hen, and even between different strains of the modern laying hen. Even when production conditions are designed to adapt the hen to a special ecological niche, there is wide variability in a number of behaviour patterns, for example, the tendency to perch on a wooden stick or a wire floor, as Dr. Faure described. So I think there is a need for basic research on such things as individual space requirements but also on the genetics of behaviour because there appear to be wide strain differences.

Perhaps we could now turn to summarising the physiological data. Dr. Petersen has stressed the importance of factors such as temperature, light, air movement and related features, also the importance of space. Dr. Beuving has talked about the possibility of measuring stress. I think these are the main points concerning the physiological responses of the fowl. Have you any comments on the most important fields for future research?

C. Beuving (The Netherlands)

I think my next step will be to try to measure cortico-sterone by hand rather than by taking samples by cannulation. We have found that we can take blood samples within half a minute. If you are quick you can do it. We have tried it and have found the same values as we obtained with cannulated birds. Perhaps it is a possibility to go on and to take blood samples in more dense populations. Another possibility is to do some work on catecholamines.

J.P. Signoret

You think it is possible in the near future to have blood sampling methods which will enable us to measure the level of corticosteroid secretion as an index of a possible stressful situation?

C. Beuving

Yes; the advantage of blood sampling methods over cannulation techniques is that you can take samples from a much larger number of birds, as many as you want.

J. Petersen (FRG)

I think there is a good possibility to do some work on climate in relation to behaviour and to physiological parameters. There is insufficient information on air velocity in this context in relation to feathering the hens. I think that is a big problem. I believe that air movement in the house does have

some influence on the feathering of hens. Another area where
there is insufficient data is that of the connection between
temperature in the houses and density of the population. Also,
investigation is needed into the genetical and phenotypical
correlation between production and behavioural traits to see
if there is antagonism which we can overcome by selection.

M. Zanforlin

I would like to go back to the question Dr. Signoret
raised about studying ethograms of the different strains. Does
anyone know how stable the commercial strains are because it
could be a never ending process to follow all the various new
strains that appear almost every year. Maybe there should be a
regulation that the producer of a new strain should specify the
ethogram and the behavioural characteristics of the animal as
well as the laying capacity and the weight gain!

W. Goldhorn (CEC)

I would like to try to put one or two things together.
Dr. Petersen said that the absolute minimum of space is about
340 cm^2 per bird because below this figure there is a dramatic
decrease in production and viability. Then we heard from
Dr. Hughes that from about 400 - 580 cm^2 there is an increase in
egg production. Does this not mean that between 400 and 580 cm^2
there is an effect, at least on welfare? Then I would like to
mention the work of Dr. Bogner. I think everyone knows his
paper on photometrical measurement of space. He found that with
a hen of 1.8 kg, the shadow is at least 450 cm^2 and for the most
primitive movements it needs at least 480 cm^2. So merely from
the physiological point of view I think 340 cm^2 is not acceptable. It should be at least 450, or even 500 cm^2 as an absolute
minimum. Is this right?

J.P. Signoret

I think there will be some overlap here with basic
behavioural requirements which we will be discussing soon. This
could be a good conclusion for stimulating the next step of our
meeting.

W. Goldhorn

Yes, because up to almost 600 cm^2 you have physiological progress which means that below 580 cm^2 there is an impact on animal welfare, maybe even on health and physiological measurements.

B.O. Hughes

I think there is some fundamental disagreement here between Dr. Petersen and myself. The literature I reviewed included some 30 papers, mostly American. Many of them had not dealt with group size and stocking densities independently. Of those which had, about 80 or 90% concluded that there was a fairly definite trend for production to increase with reduced stocking density. About 12 of them stopped at about 600 cm2; hardly any of them took any readings above that figure. So it is possible that the trend continued beyond that. I think we must assume that this is likely to be the case; there is unlikely to be a sharp threshold. The effect is likely to continue but to become less and less marked as stocking density becomes a less important feature and something else enters it instead. As far as the lower figure is concerned, 340 cm^2, I think only one study has looked at birds stocked more densely than this. Dr. A. Hill in Canada did try to house birds at 280 cm^2 but he found mortality was so high and egg production so poor that he terminated the experiment very quickly because it was obviously unacceptable for humane reasons to keep birds at such high stocking levels.

So, we have to say that there is some disagreement between Dr. Petersen and myself here; his experience from the literature is not my experience. There does appear to be a conflict of interpretation.

I.J.H. Duncan

I would like to make a comment on this debate which is taking place and link it in with what Dr. Zanforlin said earlier. Are we being realistic in thinking that we can set some limit for the hen? We have been hearing all morning about the

variability that exists between strains of hens; it is much
more likely that we would have to say, for example, that the
limit for a Ross A strain, 1980, is such and such. It is un-
realistic to expect to arrive at one figure that will cover all
strains.

B.O. Hughes

There has only been one good study done on comparing
various strains at different stocking densities. It was done
by Christmas, Harms, Rowland and others about 5 years ago. They
found in one laying year big differences between strains which
tends to support what Ian Duncan has said. When they repeated
the experiment the following year, they found big differences
between strains again but they were not consistent with the
findings of the previous year. This suggests either that the
strains vary from year to year, or that there are some other
factors coming into it producing strain/year interactions which
makes it very complex.

C.C. Brantas *(The Netherlands)*

I agree with Dr. Duncan's remark. When we are talking
about a minimum of about 360 cm^2 per bird, we must realise that
in the Netherlands a rather large number of laying hens are
kept in conditions where they have no more than 320 cm^2. Until
now these conditions are not profitable for the farmers but
they think they are profitable. I am quite sure that in the
future we agriculturists, we physiologists, we ethologists, and
so on, can make it possible to keep birds in such conditions
with profit - also in conditions of 290 cm^2 or even less. I
think the only minimum we can speak of is 1 hectare because when
laying hens are kept in totally free conditions they join up in
groups and live in a territory of about 1 ha. So the minimum is
1 ha. When we have these animals in less space we are depressing
their welfare. That is the only measure we can have - 1 ha.

J.P. Signoret

I would like to make a point here. If animals have been domesticated over a long period of time it is probably because they are animals able to undergo limitation of space without major detrimental effect. This is true of pigs, a species which I know more about than poultry. The home range for a group of female wild pigs is some 400 ha, but studies show that the principal reason for this size of territory is their search for food. If a food supply is available in a limited area through-out the whole year, this range may be markedly reduced, even though the animals are not fenced in. So, having this flexibility in space requirement has made the pig an obvious choice for domestication. Maybe the same applies to poultry. I don't know whether studies have been done in feral or wild populations, but I think this could well be a factor.

C.C. Brantas

You speak of 'detrimental effects', but what is a detrimental effect? In the eyes of the farmer it is quite a different thing from in the eyes of a welfarist. In the eyes of an animal welfarist the limitation of space, as such, is a detrimental effect.

J.P. Signoret

As a student of animal behaviour my own opinion is that there is a detrimental effect when the animal is undergoing limitation that is making it uncomfortable. If animals show exactly the same behaviour patterns when they are limited by a fence as when they are wandering freely, then it can be concluded that they have no special space requirement. There is no objective reason for saying that they are better in one condition than the other. On the other hand, there are some animals which are physically unable to undergo a limitation of their space, for example, the male roe deer. It is absolutely abnormal to confine a male roe deer in less than 15 ha of space.

C.C. Brantas

That is your explanation. There are very many personal opinions given in this regard.

J.P. Signoret

Yes, this is my opinion; it is not a conclusion of the meeting.

W. Sybesma *(The Netherlands)*

Are you not, in fact, making a plea for selection of 'fearless' animals?

J.P. Signoret

It could be possible in species such as the domestic fowl.

J.M. Faure *(France)*

I think there are two alternatives: you can allow one hectare per bird, or you can try to have birds which will manage with less than 1 ha. I think it is possible to reduce the space requirement. In giving a figure of 1 ha, Dr. Brantas was at a too low level because if you consider game cocks, then 1 ha is not enough - they will fight until one of them dies. That is an extreme case of course. I think we must work from the other end and it is possible to reduce the space requirement, or fear, of the birds. It is the only way if we want to have domestic animals, not only chickens.

M. Zanforlin

On the process of domestication, I believe that Colin Lawrence said that if you can keep a species in a close environment and still obtain reproduction then the conditions are reasonably good. Whatever the species you catch in the wild and bring into close confinement, you observe a general law. There is very high mortality for the first few generations and then productivity starts to increase as the animal adapts itself to its new environment. So, I do not think it is valid to

consider the space requirement of the species in the wild as a
basis for the requirement of the animal in captivity.

J.P. Signoret

We might find the answer in the importance of the genetic
programme in this space requirement as it appears that in some
species a genetic programme leads to such a large space requirement that it makes the animal unable to live without this,
whereas in other species this variability could be used for
animal breeders.

M. Zanforlin

Yes, but you also have the species which cannot be
domesticated.

J.P. Signoret

Yes, and their only possibility when kept in a limited
space is to die.

W. Goldhorn

I agree with you that there are strain differences and
behavioural differences and therefore the space requirements are
different. However, would you agree that the absolute minimum
should be the space of the shadow the bird throws on the ground,
plus, say, 10% to allow for movement?

I.J.H. Duncan

I think that is far too small.

J.M. Faure

You are speaking of the hen as one species, but I think
that this room is much too small for two game cocks. It is an
extreme, I agree, but you cannot specify one figure for all the
different types of hens.

R. Moss

May I remind the meeting that this is a scientific meeting designed to look at the needs of research and development. Let us forget for the time being that sooner or later someone is going to ask about a figure. Perhaps the advice that has to be given from this scientific symposium is that it is impossible to give a figure except in relation to Dr. Brantas' point of 1 ha. That, of course, would be unacceptable.

J.P. Signoret

Or to give a specific figure for a given strain from a given origin.

We must now close this discussion. I thank the people who have given papers and those who have contributed to the discussion. It has been very stimulating, even if we have not been able to arrive at a complete answer to the questions posed.

THE REGULATION OF DUSTBATHING AND OTHER BEHAVIOUR PATTERNS IN THE LAYING HEN: A LORENZIAN APPROACH

K. Vestergaard

INTRODUCTION

In discussing behaviour and welfare of the chicken and other domestic species, Wood-Gush (1973) and Duncan (1974) have stressed the significance of how the behaviour in question is regulated, that is the way of acting of the underlying motivational mechanism.

Ethologists have presented several 'models of motivation' to explain the principles of the mechanism that regulated the behaviour they were studying and which fit their results. In general, such models fail to explain all behaviour regulation in a wide range of species and of motivations. Some behaviour patterns may fit the model while others may not.

The present subject is to discuss some aspects of chicken behaviour in the light of the ideas developed by K. Lorenz (Lorenz, 1950). The Lorenzian point of view on the regulation of behaviour patterns can be illustrated by the so-called 'psycho-hydraulic' model of behaviour (Figure 1). The model predicts that without external stimulation the motivation for the behaviour in question accumulates (1) and that under such circumstances vacuum activities will eventually occur (2). Furthermore, if external stimuli are presented, performance of the behaviour pattern results in a reduction or an exhaustion of the motivation (3). In the latter case, the performance of the behaviour pattern results in a period of 'quiescence', that is the behaviour pattern will be absent for some period of time.

In the following we shall examine these three predictions in detail and their consequences with special emphasis on

R. Moss (ed.), The Laying Hen and its Environment, 101-120

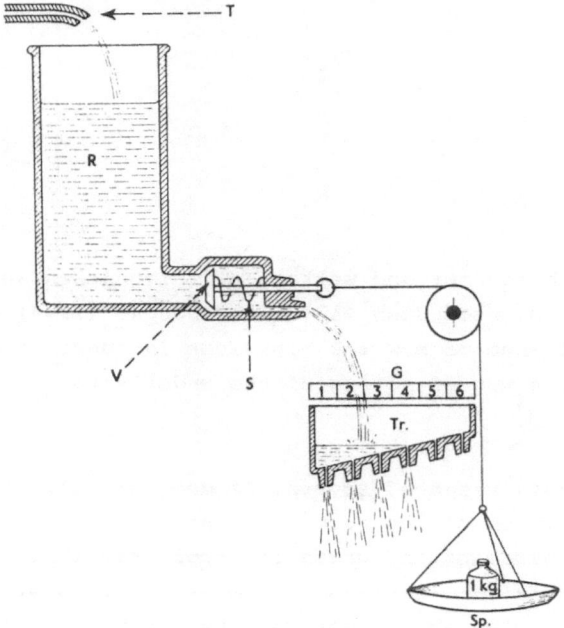

Fig. 1.   Lorenz's 'psycho-hydraulic' model of behaviour

Briefly, the model acts as follows:

1) Action specific energy (R) (represented here by water) comes from endogenous sources and fills up the reservoir.

2) Both the level of the water, that is the internal motivational state, and the weights on the pan, that is external stimuli, contribute to the opening of the cone valve (V). The cone valve represents the releasing mechanism; the inhibitory function of higher centres is represented by the spring (S).

3) Outflow from the reservoir through the graded trough (G and Tr.) represents the performance of behaviour patterns: the higher up the trough the more intensive the patterns (belonging to the behaviour in question) and the higher their thresholds.

4) It appears from the model that without external stimuli (weights on the pan) the internal motivation (water) accumulates until eventually vacuum activities occur. It also appears that after performance of behaviour, especially in the case of very strong external stimuli, the water may have flowed out. In that case no more behaviour can be elicited, no matter how strong the stimuli.

dustbathing, although whenever possible other behaviour patterns
will also be included. However, before examining the three
predictions it will be relevant to say a few words about the
dustbathing elements and their function.

THE DUSTBATHING ELEMENTS AND THEIR FUNCTION

Dustbathing in the fowl consists of bill-raking, scratching
(with the leg), vertical wing shaking (with one or both wings),
head rubbing and side rubbing (Kruijt, 1964; Nice, 1962). It is
always terminated by one or several body/wing shakes which may
also occur under other circumstances. During dustbathing each
of the elements is repeated several times but they always occur
in the order mentioned above (Nielsen, 1979). During dustbathing
dust is thrown into the plumage and at the end of the bout dust
is expelled from the plumage by the body/wing shaking movements.
Dustbathing in gallinaceous fowls probably functions to remove
old oils (originating from the uropygial gland) from the plumage
(Borchelt and Duncan, 1974). It has been shown that dustbathing
in quails improved barb alignment, made the down more fluffy and
dry and reduced the amount of dandruff on the feathers (Healey
and Thomas, 1973).

THE PREDICTIONS OF THE PSYCHO-HYDRAULIC MODEL

Let us now examine the predictions arising from the psycho-
hydraulic model by answering the following three questions:

1. Is there evidence of an accumulation effect?

2. Have vacuum activities been observed?

3. Is there evidence of reduced motivation or quiescence
    after performance of behaviour patterns?

1. Recent experiments (Vestergaard, 1980) have shown that
deprivation in birds kept on wire floors for from 5 to 101 hours
resulted in a gradual increase in the readiness to perform dustbathing (Figures 2 and 3), and of the length of the dustbathing

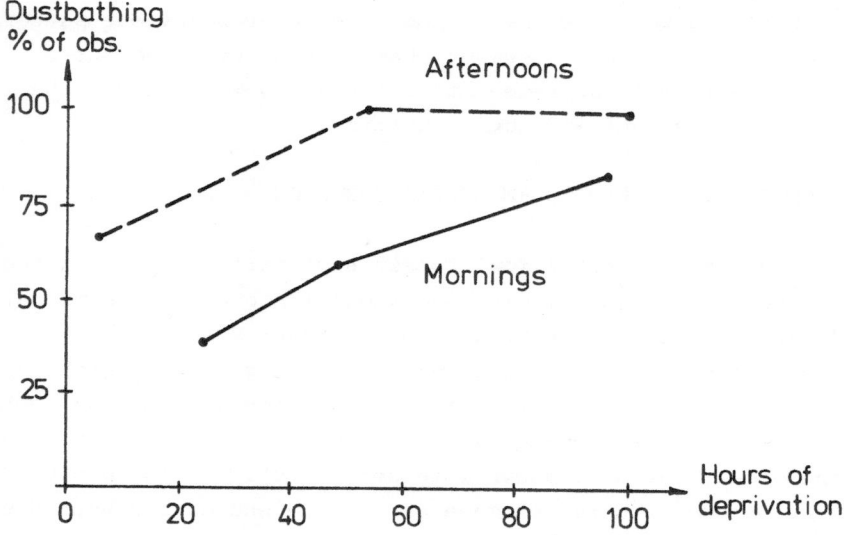

Fig. 2. Percentage initiation of dustbathing within one hour of access to
dust. The mean number of hens which initiate dustbathing increases
with length of deprivation. The numbers are smaller in the mornings
than in the afternoons (From Vestergaard, 1980).

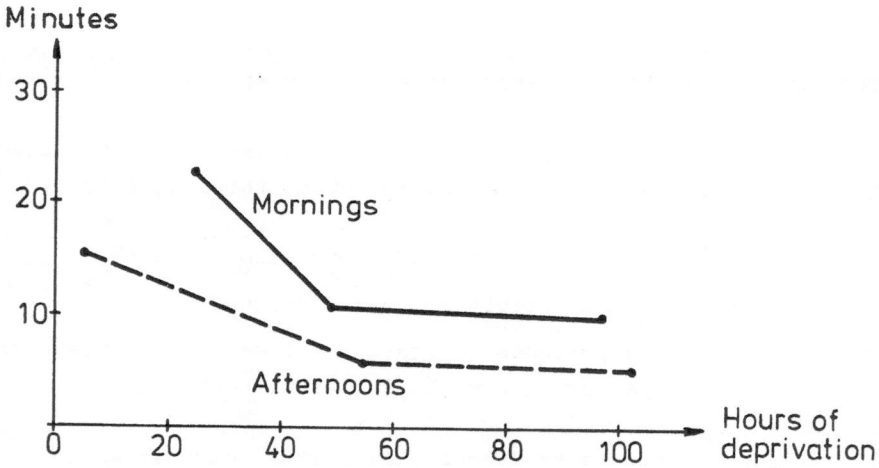

Fig. 3. Latency of dustbathing as a function of deprivation. Latency
decreases with length of deprivation and the values are smaller
in the afternoons than in the mornings (From Vestergaard, 1980).

bouts (Figure 4). Similar results have been obtained for quails
of different species (Borchelt, 1975; Benson and Schein, 1965).
Accumulation effects have also been recorded in laying hens
after a three weeks stay in conventional battery cages (Wennrich,
1977). Apart from dustbathing, Wennrich also recorded accumulation effects of ground scratching, wing flapping, vertical wing
stretching, body/wing shaking and tail shaking (Figure 5).

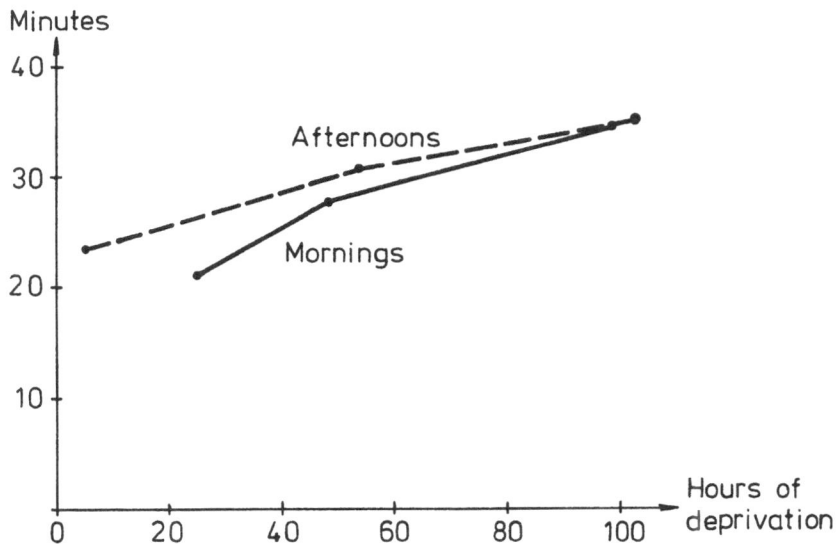

Fig. 4. Duration of dustbathing after deprivation. The duration increases
with length of deprivation. There is no significant difference in
duration between morning and afternoon tests (From Vestergaard, 1980).

FREQUENCY OF

THE BEHAVIOUR

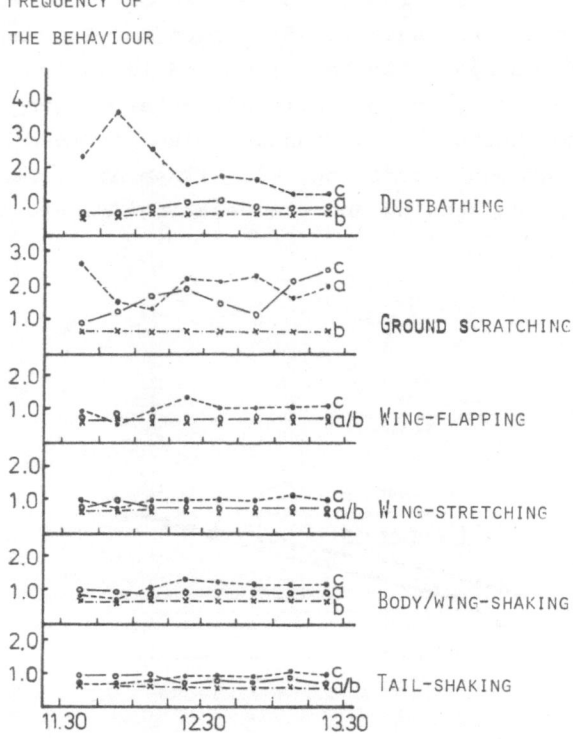

Fig. 5.   Interactions between housing conditions and observation time of six behaviour patterns observed before (A), during (B) and after (C) a three week stay in battery cages. Before and after the stay in the cages the hens were kept in litter pens (From Wennrich, 1977).

Both a possible rise in internal motivation and an accumulation of oil in the feathers (Borchelt and Duncan, 1974) together with other sources of external stimulation, may contribute to the increased tendency to perform dustbathing. However, oil gland removal has failed to reduce the increase in the dustbathing tendency after deprivation (Nielsen, 1979).

2.   Hens kept on wire floors frequently dustbathe although at a lower rate than those that are provided with litter (Figure 6). Similar observations have been made with hens in cages

(Black and Hughes, 1974; Fölsch and Huber, 1977; Wennrich, 1977).
Vacuum dustbathing on wire floors probably needs more than 100
hours of deprivation in order to occur because it was not
observed at all during my deprivation study mentioned above.
In that study each hen was kept on wire at each deprivation
level more than 15 times. Thus learning seems to play a minor
role. Quails have also been recorded as performing dustbathing
on wire floors in a social facilitation situation (Healey and
Thomas, 1973). Apart from dustbathing, vacuum nest building
activities have also been reported in caged laying hens
(Wood-Gush, 1972).

GROUP	FREQUENCY OF DUSTBATHING	
SL	1.91	DEEP LITTER PENS
LL	2.44	
SW	1.09	SLOPING WIRE FLOORS
LW	0.81	

Fig. 6. Dustbathing in deep litter pens and on sloping wire floors.
Dustbathing occurs in both environments but the frequencies
are lower on the wire floors than in the litter pens.

SL = Small deep litter pen   LL = Large deep litter pen
SW = Small wire floor   LW = Large wire floor

(Vestergaard, unpublished)

3.   It appears from Figure 2 that fewer hens perform dustbathing after recent access to dust than those deprived for
longer periods of time. When provided with litter continually,
as a mean, the hens only performed one bout of dustbathing
every second day, and more than one bout per day was very rare.

INFLUENCE OF DEPRIVATION ON DIURNAL RHYTHMS

The psycho-hydraulic model does not directly predict
changes in diurnal rhythms, but if deprivation has any
accumulation effect at all on a behaviour pattern we should

expect such disturbances. This is simply to say that when an
animal is strongly motivated to perform a behaviour pattern,
the presentation of the relevant stimuli at almost any time of
the day invariably releases that behaviour pattern. This is
true for dustbathing since it is evident from Figures 2, 3 and 4
that with increasing deprivation time the birds, at an increasing
rate, perform dustbathing in the morning although this is clearly
outside the normal time of day for that activity (Figure 7).
Similar effects have been recorded by Wennrich (1977) for dustbathing, ground scratching, wing flapping, body/wing shaking
and tail shaking (see Figure 5).

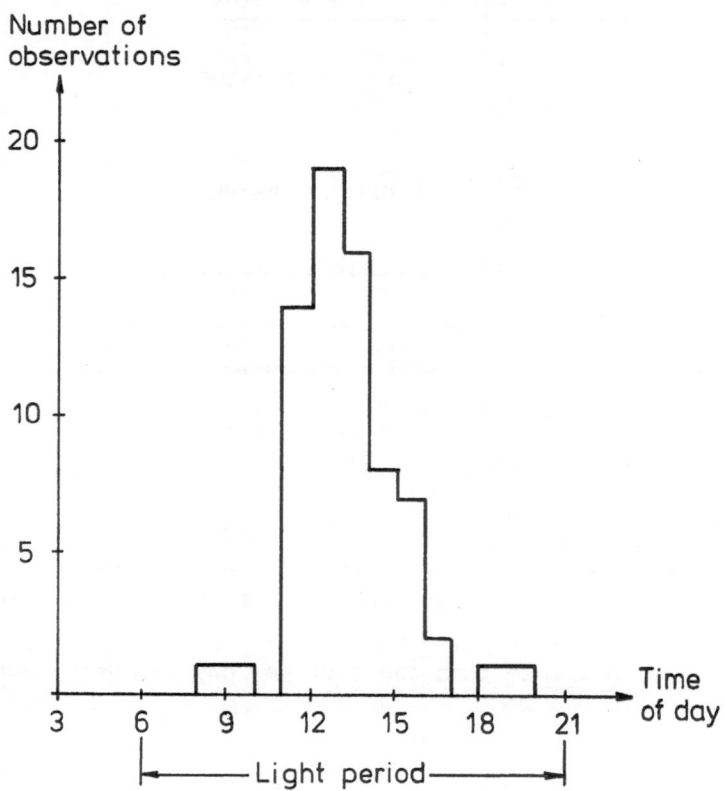

Fig. 7. Diurnal rhythm in the initiation of dustbathing during free
access to dust (From Vestergaard, 1980).

CONSEQUENCES OF THWARTING HIGHLY MOTIVATED BEHAVIOUR

There have been only a few studies concerned with such consequences; most of these deal with feeding and pre-laying behaviour which do not fit the psycho-hydraulic model. In a study of hens on wire floors (unpublished) I provided the birds with a dust box and found a marked decrease in the amount of aggressive pecking and threats (Figure 8). The effect only occurred in one of the two flocks studied, namely the one which learned to use the dust box for dustbathing. The other flock hardly used the dust box with the exception that the birds pecked in the dust from the outside.

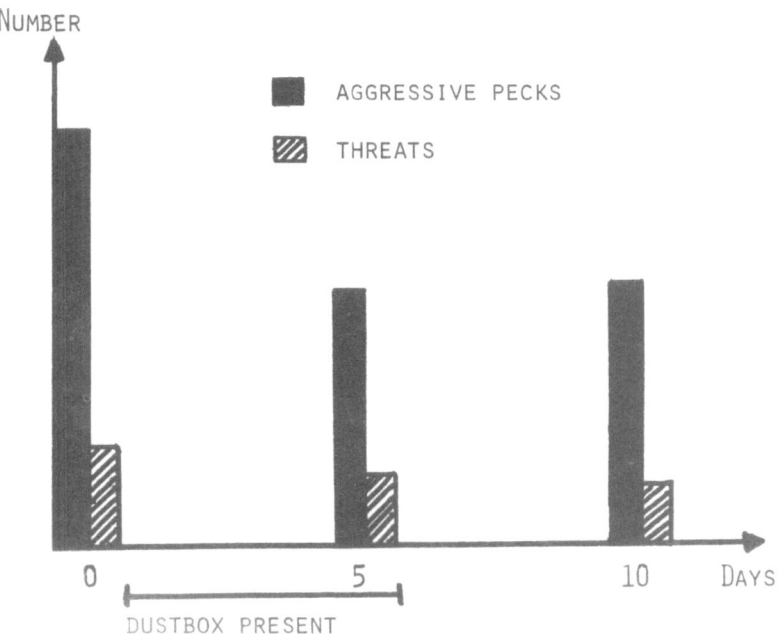

Fig. 8.   Frequency of aggressive pecking and threats in a wire floor group of layers before, during and after the presence of a dus⁺ ʰᵒx. The number of pecks decreased during access to du⁻   ⅃ remained low five days later (Vestergaard, unpublished).

Increased aggression and other signs of frustration have been found during thwarting of feeding in hungry hens (Duncan, 1974). Under certain circumstances, some strains may also react with increased aggression during pre-laying periods when no litter is available for nest building purposes (Hughes, 1979). In both situations, stereotyped pacing behaviour may occur (Wood-Gush and Gilbert, 1975). The pacing may be a sign of fear since it is reduced by administration of pacitracin (Duncan and Wood-Gush, 1974).

Thwarting of wing flapping in the cages can cause a reduction in bone strength of the humeri. This in turn may result in broken bones during loading and transport (Simonsen and Vestergaard, 1978).

DISCUSSION AND CONCLUSIONS

Strong evidence has been presented for accumulation effects during deprivation of dustbathing and for a quiescent period after the performance. Conclusive evidence of vacuum dustbathing is available from several studies. It may be argued that no real vacuum dustbathing (or nest building) has been recorded since it is hard to exclude any possibility of stimulation arising from the environment. However, as pointed out by Manning (1972):

"In a strict sense it is impossible ever to designate an activity as vacuum; some external stimulus, no matter how minute, may be there. Nevertheless the term does draw attention to the extreme lowering of threshold which can occur."

The role of external stimuli from the integument and the plumage is still open to question, although it seems that accumulation of oil originating from the uropygial gland is of minor significance. If such stimulation increases with deprivation it will, alone or in combination with an internal motivation mechanism, lower the threshold. Whatever the

mechanism, the result for the bird is the same: a strong urge
to perform this behaviour when deprived of dust.

According to Duncan (1974) frustration is almost
inevitable if a behaviour is regulated in accordance with the
psycho-hydraulic model (and no relevant stimuli are present).
He also stresses that the occurrence of vacuum activities would
have to be anticipated and that sufficient space would have to
be allowed for their performance in order to prevent further
frustration. However, it is still open to question whether or
not vacuum performance reduces the motivation.

The scanty evidence presented here of frustration due to
dust deprivation clearly needs to be confirmed and examined in
greater detail before firm conclusions can be drawn about the
influence of dust deprivation on the psychical well-being of
the hen.

In addition to the psychical effects of dust deprivation
there may also be some physical changes in the integument and
plumage, some of which have already been mentioned. The link
of such changes with well-being is still unknown.

Unlike dustbathing and wing flapping, for example, the
urge to perform nesting activities rises to a high level each
day of laying and then declines again within a few hours.
However, during that time accumulation may occur in the absence
of stimuli (straw). Vacuum nestbuilding occurs in this
situation and, as with vacuum dustbathing, this indicates an
extreme lowering of the threshold.

In conclusion, it seems that some innate behaviour
patterns, especially those of dustbathing, fit the psycho-
hydraulic model of motivation to a certain extent. When such
behaviour patterns are being thwarted frustration can be
expected.

112

REFERENCES

Benson, B.N. and Schein, M.W., 1965. Factors influencing dustbathing in
    Coturnix quail. (Motion picture). Amer. Zool. Abstr., 5, 196.

Black, A.J. and Hughes, B.O., 1974. Patterns of comfort behaviour and
    activity in domestic fowls: A comparison between cages and pens.
    Br. vet. J., 130, 23-33.

Borchelt, P.L., 1975. The organisation of dustbathing components in
    Bobwhite quail (Colinus virginianus). Behaviour, 53, 217-237.

Borchelt, P.L. and Duncan, L., 1974. Dustbathing and feather lipid in
    Bobwhite quail (Colinus virginianus). Condor, 76, 471-472.

Duncan, I.J.H., 1974. A scientific assessment of welfare. Behav. Biol.,
    8, 109-114.

Duncan, I.J.H. and Wood-Gush, D.G.M., 1974. The effect of a rauwolfia
    tranquillizer on stereotyped movements in frustrated domestic fowl.
    Appl. Anim. Ethol., 6, 67-76.

Fölsch, D. and Huber, F., 1977. Bewegungsaktivität und lautausserung im
    tagesrhytmus bei hühnern. KTBL-schrift (Darmstadt), 223, 99-114.

Healey, W.M. and Thomas, J.W., 1973. Effects of dusting on plumage of
    Japanese quail. Wilson Bull., 85, 442-448.

Hughes, B.O., 1979. Aggressive behaviour and its relation to oviposition
    in the domestic fowl. Appl. Anim. Ethol., 5, 85-93.

Kruijt, J.P., 1964. Ontogeny of social behaviour in Burmese Red Junglefowl,
    (Gallus gallus spadiceus). Behaviour Suppl., 12, 1-201.

Lorenz, K.Z., 1950. The comparative method in studying innate behaviour
    patterns. Symp. Soc. exp. Biol., 4, 221-268.

Manning, A., 1972. An introduction to animal behaviour. 2nd ed. London, 1-294.

Nice, M.M., 1962. Development of behaviour in precocial birds. Trans.
    Linn. Soc. N.Y., 8, 1-211.

Nielsen, G.N., 1979. En deskriptiv og eksperimentel undersøgelse af
    støvbadningsadfærden hos tamhøne (Gallus domesticus). Speciale-
    rapport, University of Copenhagen. 1-87.

Vestergaard, K., 1980. Dustbathing in the domestic hen: diurnal rhythm
    and dust deprivation. Appl. Anim. Ethol. (In press).

Simonsen, H.B. and Vestergaard, K., 1978. Battery cages as the cause of
    environmental and behavioural dependent diseases. Nord. Vet. Med.,
    30, 241-252.

Wennrich, G., 1977. Zum nachweis eines 'Triebstaus' bei Hausbennen
(*Gallus gallus var. domesticus*). KTBL-schrift (Darmstadt), 223,
115-129.

Wood-Gush, D.G.M., 1972. Strain differences in response to suboptimal
stimuli in the fowl. Anim. Behav., 20, 72-76.

Wood-Gush, D.G.M., 1973. Animal welfare in modern agriculture. Br. vet.
J., 129, 167-174.

Wood-Gush, D.G.M. and Gilbert, A.B., 1975. The physiological basis of a
behaviour pattern in the domestic hen. Symp. Zool. Soc., Lond.,
35, 277-306.

DISCUSSION

H.C. Adler   *(Denmark)*

In order to underline the significance of this Session I would like just to quote a few words from the European Convention for the Protection of Animals kept for Farming Purposes. From Article III:

> "Animals shall be housed and provided with food, water and care in a manner which is appropriate to their physiological and ethological needs."

From Article IV:

> "Where an animal is tethered or confined it shall be given the space appropriate to its physiological and ethological needs."

From Article V:

> "The lighting, temperature, humidity, air circulation, ventilation and other environmental conditions, such as gas concentration and noise intensity, shall conform to its physiological and ethological needs."

Now are there any questions or comments?

G. Martin   *(FRG)*

Dr. Vestergaard, you mentioned this vacuum activity. Did you also notice the birds in cages trying to get their feet into the trough with the purpose of indulging in sand bathing activity?

K. Vestergaard   *(Denmark)*

Yes, I have seen this kind of activity.

G. Martin

Do you think that the bird really suffers deprivation from the lack of dust, litter or straw?

K. Vestergaard

From the evidence we have so far it would seem that the birds really do miss those things; it may cause frustration. I compared wire floors with deep litter; when the birds had a dust box there was a reduction in aggression. So it points to the fact that there might be some frustration effect without dust. Also, when we compared wire floors with litter we generally found a higher aggression level in the wire floors.

G. Martin

Yes, this is my experience too. You found the same thing Dr. Brantas, I think?

C.C. Brantas   *(The Netherlands)*

Yes.

K. Vestergaard

There has been a lot of debate in Denmark about these results because I have never published them. The experiment has been repeated by someone else at our Department and he found the same. That was under very controlled conditions; he had two large flocks, 120 hens, and he observed the aggression around the feeding troughs.

G. Martin

The fact that the bird will subject itself to such contortions to try to get into the trough shows that the internal drive is very high for this movement.

K. Vestergaard

Yes, in the litter groups I never observed this kind of reaction to the feed so it seems reasonable to conclude that the feed is a lower stimulus for dust bathing than the litter. However, in the case of dust bathing at the feeding trough, you cannot say that there is no other stimulation because it might be that the feeding trough elicits the behaviour. But then it is difficult for the bird to continue; it cannot maintain the squatting position for very long. In that case you cannot say that it is a vacuum activity, but I have seen dust bathing in many other places on the wire floor and that is more likely to be a vacuum activity.

J.P. Signoret (France)

Are there indications of inter-strain variations in dust bathing frequency? The very high variability reported this morning might suggest either a variability or something highly fixed, as suggested by the Lorenzian model, which is mainly from a genetic programme origin. It would be interesting to know whether this is a very general and constant drive in poultry, or whether it is subject to strain variations.

K. Vestergaard

I really don't know, but at least it seems that we are dealing with fixed action patterns. We have described all these patterns in detail; they are very fixed, very stereotyped. However, according to some authors, for example, the text book by Alcares, it seems that maintenance behaviour, grooming and so on, is the most resistant to selection for behaviour patterns.

P.M. Schenk (The Netherlands)

Perhaps I can give some additional information on strain differences. We are also working on dust bathing in Junglefowl and we are trying to compare them with an old Dutch breed. All the same elements are present in both but the Dutch breed birds have a slower tempo.

K. Vestergaard

So it is the domesticated birds that have the slower
movements?

P.M. Schenk

Yes.

I.J.H. Duncan   *(UK)*

I have one or two comments on dust bathing. I think your
evidence for frustration is rather slight.

K. Vestergaard

I agree.

I.J.H. Duncan

It is certainly true that aggression can be a result of
frustration but I don't think you can necessarily argue the
opposite way by saying that if aggression is decreased then
the birds must have been frustrated beforehand. That is what
your results showed, that you had a particular level of
aggression and when you allowed the birds to dust bathe that
aggression decreased. That can simply be a matter of time
availability and the amount of moving about of the birds, and
coming in contact with other birds. If it really was a frustration effect one would expect aggression to rise up afterwards
when you take the dust bath away. I think that is an essential
part of showing that it is a frustration effect.

K. Vestergaard

I agree that there are difficulties of interpretation of
the results, but I measured the aggression at the feeding troughs
and I counted the number of birds in the area and that number
was the same during all these observations. So I had the same
number of birds but in some cases some of the birds might have
performed dust bathing whereas in other cases they did not.

I.J.H. Duncan

There is another point I would like to make. You used the example of there being a diurnal rhythm of dust bathing. We know, for example, that birds do show a diurnal rhythm of the important maintenance activities such as feeding and nesting. Once again, it may be that at times when these other needs were satisfied the attention of the birds switched to things which were constantly present such as preening or dust bathing. If the bird was feeding and nesting at the beginning of the day and feeding again in the latter part of the day, that might leave the middle part available for things of less importance, not of more importance.

K. Vestergaard

Yes, that is true, but the main thing is that it comes early in the morning after deprivation, that is to say that the other elements in the competition for the attention of the hen are being pressed aside. Also, light intensity may contribute to the regulation of the diurnal rhythm but in this case we had a very strictly controlled environment and the same light intensity all day.

I.J.H. Duncan

My final point is this: how do you explain the fact that one of your flocks 'did not learn to dust bathe when given the dust box' if it really is a fixed action pattern?

K. Vestergaard

It is very difficult to explain. The two flocks differed in respect to density. In the flock which did not perform dust bathing the density was half that of the other one. There were only seven hens/m^2 as opposed to 14 hens/m^2 in the flock which did perform dust bathing. I think that what we are dealing with here is simply the fear of novel objects. They did not have the dust box long enough and they could easily avoid it because they had plenty of space.

J. Fris Jensen *(Denmark)*

Your results were obtained on groups of 120 birds on a kind of Pennsylvania system. What do you think is the effect of such a large group on the desire to dust bathe, compared with 3 or 4 birds in a cage? Can you compare the two situations?

K. Vestergaard

There were some accumulation differences at the start.

J. Fris Jensen

You said the birds were on wire in both cases but it is two different types of wire systems.

K. Vestergaard

Yes, but at least in both cases dust was absent; that is the common denominator. I think the conditions were almost the same. Of course, it is more difficult to perform vacuum dust bathing in a cage because it involves some violent movements and the birds are very close together.

J. Petersen *(FRG)*

You showed us some pictures of feathers from quails before and after dust bathing. Were these feathers from the same bird and how widespread was the effect?

K. Vestergaard

It was not my own work, it was from another study and I cannot remember the details, but there were several birds in it. The feathers were taken from the same part of the bird each time.

B.O. Hughes *(UK)*

Leading on from Dr. Fris Jensen's point, when birds do dust bathe in cages they generally do it in the front of the cage in contact with the back of the food trough. Do you think that when there is no dust available, contact with some solid object is an important triggering stimulus?

K. Vestergaard

I think a solid surface is better than wire and probably the feed is better than a solid surface.

B.O. Hughes

So you think it is the feed rather than the trough?

K. Vestergaard

It might be that.

B.O. Hughes

In our cages they cannot squat on the floor and make the movements and rake in the food at the same time, it is impossible because of the distance down into the trough.

K. Vestergaard

That is the same as in the wire floors I was studying where the troughs were 20 cm above the floor. The most common place to perform dust bathing was at the covering board and the wooden supports of the wire floor. The birds pecked at this board and then they performed dust bathing.

M. Zanforlin   *(Italy)*

Is there any particular age at which they develop this dust bathing behaviour pattern?

K. Vestergaard

It begins very early, at about one week. It is very interesting that the different innate patterns appear in the same order during development as during a normal bout of dust bathing.

H.C. Adler

Thank you very much ladies and gentlemen for a very interesting discussion.

ESSENTIAL BEHAVIOURAL NEEDS

D.W. Fölsch

INTRODUCTION

The purpose of this paper is to give some results of Swiss research projects. The following examples are only parts of complex projects which were carried out at the Veterinary Surgery Clinic of the University of Zurich and at the Institute for Animal Production at the Federal Institute of Technology.

During the period between 1972 and 1975 at the University of Zurich we undertook the task of investigating the keeping systems currently in use for hens. These keeping systems were, of course, on a smaller scale than those in commercial farming practice. The systems were: free range, deep litter, wire netting floor and battery cages. With the exception of free range, all the systems were duplicated. In the three duplicated indoor systems the hens in one section had been reared on deep litter and the others in cages. Technical data concerning the experiment and the commercial effect of the behaviour are described by Fölsch et al., 1977.

From 1977 onwards, at the Federal Institute of Technology in Zurich, we concentrated on the evaluation of behaviour and health of laying hens in deep litter in comparison with those in battery cages, and the influence of different stocking densities.

METHOD OF INVESTIGATION AND EVALUATION

A quantitative ethogram for each system was carried out by the multi-moment technique. Each animal was observed four times per hour as indicated by a timer. The protocol of the positions of behaviour was registered for each hour of the

R. Moss (ed.), The Laying Hen and its Environment, 121-147

The sum of battery reared hens in deep litter, wire netting floor and cages

▓▓▓▓ Dimlight   ○════════ Dec. 1974, 1 % = 54.7 observations/hour
▓▓▓▓ Darkness   ●•••••••• April 1975,1 % = 45.6 observations/hour

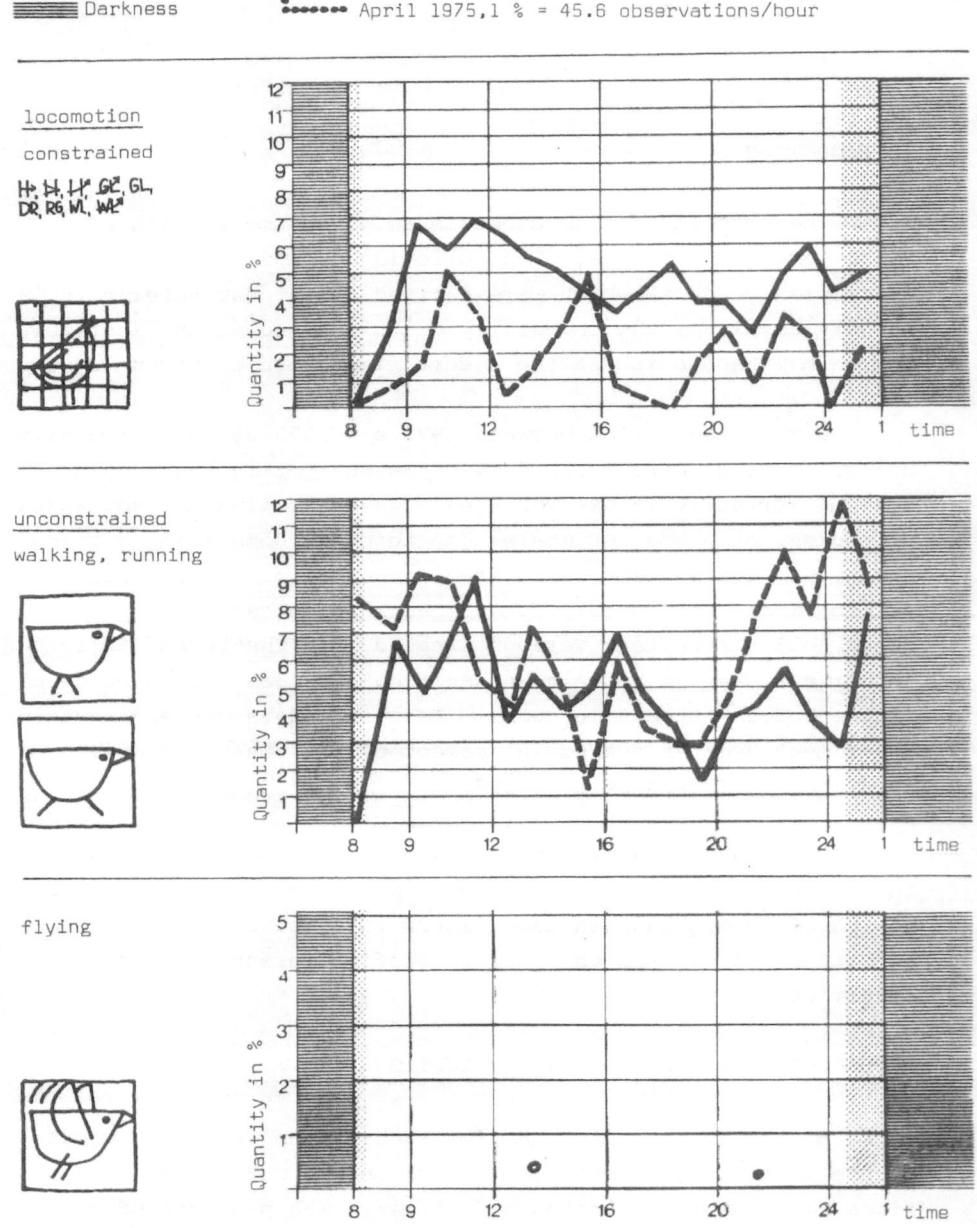

Fig. 1.   Quantities of behaviour patterns during the light hours in
          two different seasons (from: Fölsch, 1980)

17 light hours per day. A more detailed description of the observation technique as well as the results and a critical examination of the keeping systems will be published later in this year by Fölsch and Vestergaard, 1980.

RESULTS

The observed behaviour patterns are quantified and also demonstrated as a percentage of all actions, each keeping system compared with the others. However, the quantified units only give the correct information when the quality of the environment and of the behaviour of the bird are properly described. Where there is restricted floor space and wire as the environment in cages, or wire or plastic as floor materials with the wire netting floor, or Pennsylvania systems, behaviour is likely to be modified.

Example 1: Locomotion

The diurnal course of constrained and unconstrained locomotion is shown in Figure 1.

The influence of the rearing system is shown in Figure 2, it is particularly noteworthy for the hens in the two units with battery cages: hens reared in cages from the first day of life until the eighteenth week and afterwards kept in cages, 8.3%, and hens reared on deep litter and subsequently kept in cages, 6.9%. The difference is significant (P < 0.05). The influence of space and floor material is also indicated and marked.

124

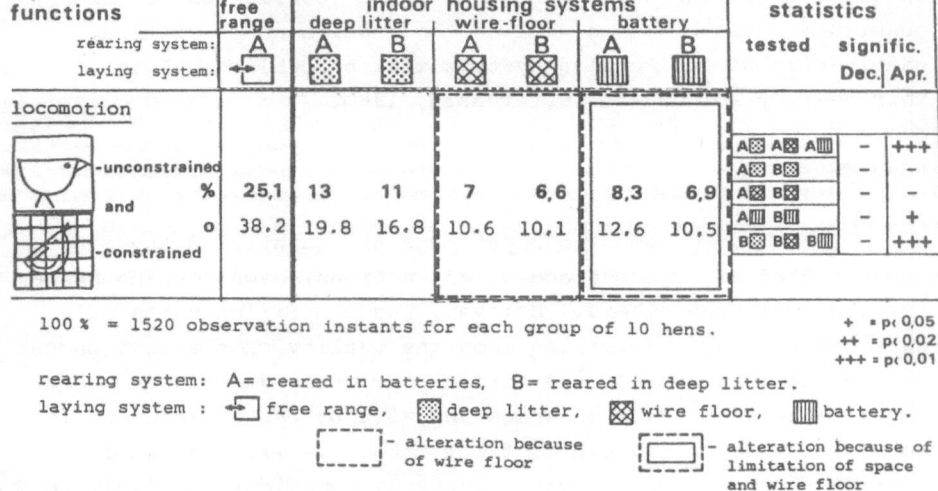

functions	free range	indoor housing systems						statistics		
		deep litter		wire-floor		battery		tested	signific.	
rearing system:	A	A	B	A	B	A	B		Dec.	Apr.
laying system:										
locomotion								A A A	–	+++
								A B	–	–
-unconstrained %	25,1	13	11	7	6,6	8,3	6,9	A B	–	–
and								A B	–	+
-constrained o	38.2	19.8	16.8	10.6	10,1	12,6	10,5	B B B	–	+++

100 % = 1520 observation instants for each group of 10 hens.

+ = p< 0,05
++ = p< 0,02
+++ = p< 0,01

rearing system: A= reared in batteries, B= reared in deep litter.
laying system : ⬒ free range, ▨ deep litter, ⊠ wire floor, ▥ battery.

⌐‒‒⌐ - alteration because of wire floor

⌐‒‒⌐ - alteration because of limitation of space and wire floor

Fig. 2. Details of observation periods (o) in % for December/April
for each housing system (from: Fölsch, 1980)

Example 2: Nesting behaviour

The behaviour of hens before, during and after egg laying
can be separated into phases for the purpose of quantifying the
time spent on each phase; they are:

Phase 1: Separation from the group, looking for nest,
nest inspection.

Phase 2: Entering the nest, behaviour patterns of nest
building, sitting quietly with eventual circular
movements of the body.

Phase 3: Bringing up the thorax to the position which I
call penguin position, laying the egg.

Phase 4: Sitting down on the nest, rolling the egg under
the body, resting quietly.

Results

    In the deep litter system with nests, hens remain in the
nest for 54 min, on average (Phases 2 - 4). The sequence of the
behavioural steps is fairly normal.

    The hens in cages, however, have no nest. These hens
spend an average of only 17 min (3 hens/cage; Phases 2 - 4) on
similar or comparable activities. In the remaining time from
60 min before, until 30 min after egg laying, the hens in cages
are upright moving backwards and forwards including pacing
(Table 1).

TABLE 1

INFLUENCE OF STOCKING DENSITY FOR HENS IN CAGES, 60 MIN BEFORE UNTIL
30 MIN AFTER EGG LAYING (duration min)

| Per cage | Lying | | Upright | | n |
| | before after | | before after | | hens/dropped eggs |
	egg drop		egg drop		
2 hens	19.6	0.9	37.6↓	29.1	14/20
3 hens	17.5 ↑	0.3	38.7↓	29.6	15/20
4 hens	12.4 ↑	0.1	44.8	29.9	14/23

    The tendency to stay for a longer time increases with
stocking density. Such behaviour patterns towards a non-existing
nest have to be called pseudo nesting behaviour in our observed
White Leghorns (Shaver and HNL, LSL).

Example 3: Mutual disturbances

    Mutual disturbances happen mostly through movement,
especially at the food trough. They are of shorter duration in
deep litter and occur less often compared with hens in battery
cages, except when 2 hens are kept per cage.

TABLE 2

MUTUAL DISTURBANCES: INFLUENCE OF HOUSING SYSTEM

	Duration (average)	Quantity (average)	Relation to all activities per light hour
Deep litter	1 min	7 times	(< 1%)
Cages	2 min	57 times	(4%)

TABLE 3

MUTUAL DISTURBANCES: INFLUENCE OF DIFFERENT STOCKING DENSITIES IN DEEP LITTER AND CAGES

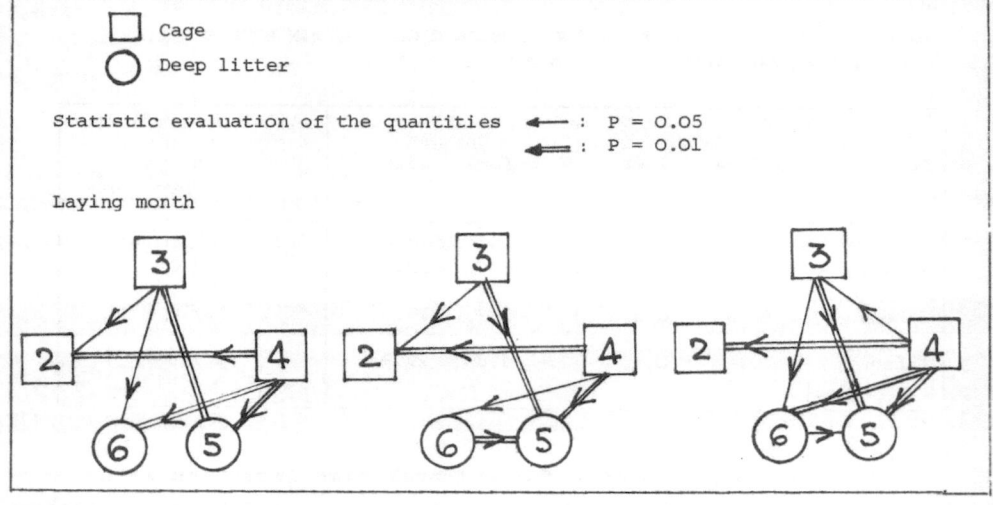

TABLE 4

MUTUAL DISTURBANCES: INFLUENCE OF DIFFERENT STOCKING DENSITY IN CAGES

Quantities during the light hours			
Hens per cage	2	3	4
Quantity	52	377	606
Ranking (Ri)	21	61	89
Relation	1 :	3 :	4

Sonogram:

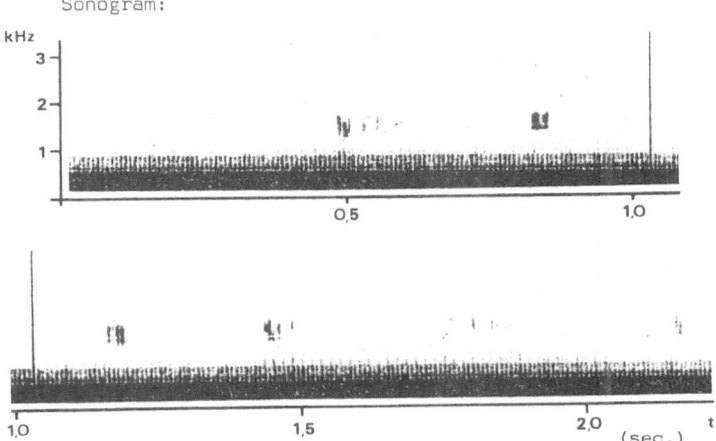

Course during light hours

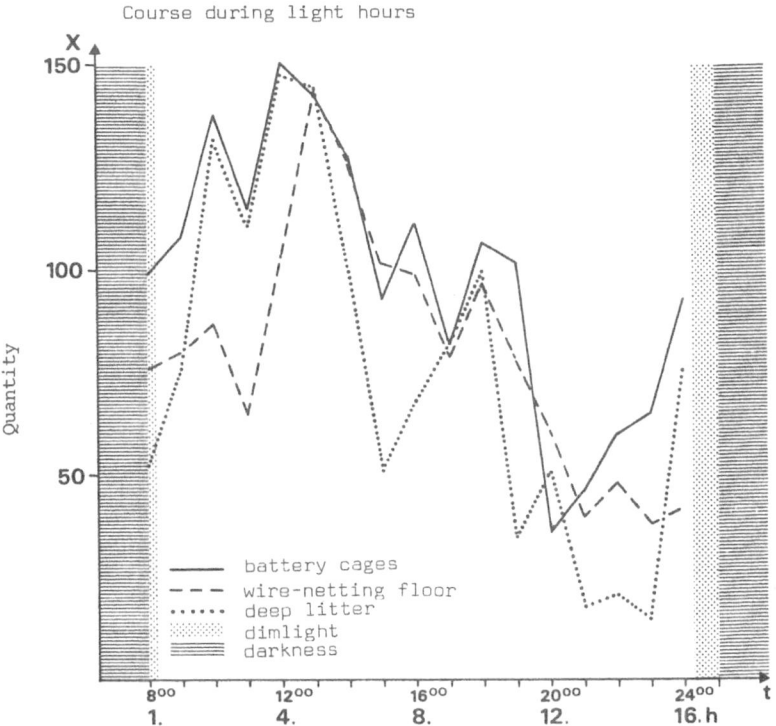

Fig. 3. Egg laying call (Gakeln) (from: Huber and Fölsch, 1978)

128

In addition to the behaviour patterns which can be observed, I would like to add some examples of the acoustic behaviour of hens (Figures 3, 4 and 5). The techniques of some sonograms are described by Huber and Fölsch (1978).

The friendly call = (freundliche Rangordnungslaut) = Ku-call (Konishi, 1963) (Figure 4).

The different levels are self-evident: hens on deep litter make the highest quantity of 'friendly' calls, then hens on wire netting floor and finally those in cages (P < 0.01).

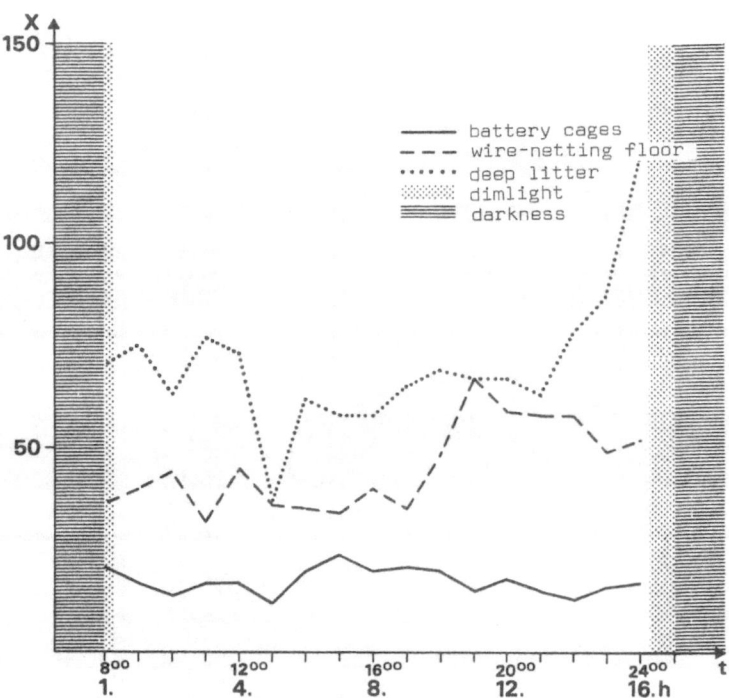

Fig. 4. Friendly call (freundliche Rangordnungslaut)
(from: Huber and Fölsch, 1978)

The dominant call = (herrischer Rangordnungslaut)) (Figure 5)

This particular call is expressed more commonly by hens in batteries as opposed to the other systems.

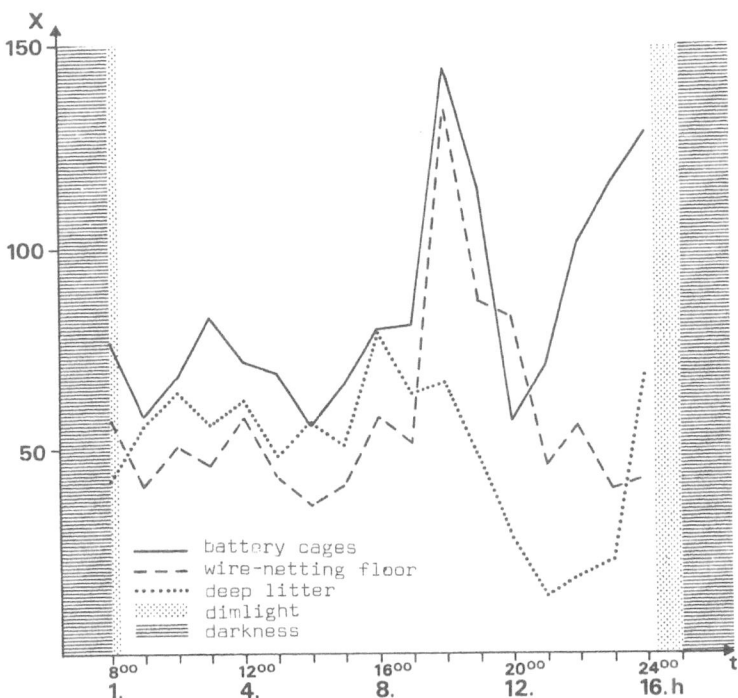

Fig. 5. Dominant call (herrischer Rangordnungslaut)
(from: Huber and Fölsch, 1978)

As with the Ku-call, the differences shown in the quantities of these calls between the three systems should be considered critically in our assessment when comparing the systems. Vocal expression can be regarded as a very positive indicator.

Apart from quantifying behaviour patterns, it might be useful to evaluate also the effect of the inanimate and living

environment of hens, for example the effect of keeping systems
and the hen population in measuring the damage to feathers and
plumage.

A practical and very exact method of testing plumage
damage produced by technopathological changes is described by
Burckhardt et al., 1979. Some facts and figures from that
publication are summarised below. In order to make a comparison
with the results obtained at our research centre, we included in
our experiments a nearby commercial egg farm and the local
cantonal agricultural school.

Quantitative examinations were made of:

1) Different rearing systems (deep litter versus battery)

2) Different housing systems (free range, deep litter,
   wire netting floor and battery cages).

The following criteria were considered:

Broken feather tips
Degree of damage to the vane (breakages in the webbing)
Air resistance of single feathers
Degree of feather loss.

The statistically significant results show that:

The degree of feather loss depends on the housing system;
wire netting floor and battery systems have a disastrous
effect on the plumage.

Damage to the single feather (broken tips, fraying and
air resistance) is dependent on the housing system.

Damage to the single feather depends to some extent also
on the rearing system. The flight feathers (remiges)
which normally lie close to the hen's body showed damage
associated with the rearing system in all the intensive
housing systems. On the other hand, damage to the tail
feathers (retrices) which sustained the most severe

injuries under battery conditions, was not dependent on the rearing system.

The degree of feather loss in battery cages depends additionally on:

1) The stocking density (P < 0.0001)
2) The height of the tiers (P < 0.003)
3) The age of the hens

TABLE 5

HENS' FEATHER LOSS IN CAGES, INFLUENCE OF STOCKING DENSITY

Per cage	Average degree of feather loss(cm^2)		n
	Laying month		
	7.	12.	
2 hens	25.7$_\downarrow$	81.8$_\downarrow$	30
3 hens	45.7$_\downarrow$	157.3$_\downarrow$	36
4 hens	55.7	231.0	44

CONCLUSION

The behaviour patterns of individual animals serve as indicators for us - behaviour is a basis for a critical examination of keeping systems. By quantifying the same qualitative patterns and symptoms, it is possible to compare different keeping systems.

The study of ethology, as applied to farm animals and birds, is a young but growing scientific discipline and embraces both applied animal ethology and veterinary ethology. The results of ethological studies are becoming more and more important in interpreting the reaction of the animal to its total environment. In such a way bridges of better understanding are being built between natural and philosophical sciences.

REFERENCES*

Burckhardt, Ch., Fölsch, D.W. and Scheifele, U., 1979. Das Gefieder des
    Huhnes. Abbild des Tieres und seiner Haltung. Tierhaltung, Vol. 9,
    Birkhäuser-Verlag, Basel, Boston, Stuttgart.

Fölsch, D.W., Niederer, Chr., Burckhardt, Ch. and Zimmermann, R., 1977.
    Untersuchungen von Legehennenhybriden unterschiedlicher Aufzucht in
    verschiedenen Haltungssystemen während einer Legeperiode von 14
    Monaten; Wirtschaftlich relevante Aspekte. Tierhaltung, Vol. 1,
    Birkhäuser Verlag, Basel, Boston, Stuttgart.

Fölsch, D.W., 1980. The behaviour of hens reared and kept in different
    housing systems. In: Fölsch and Vestergaard: Das Verhalten von
    Legehennen. Einfluss von Aufzucht und Haltung. Tierhaltung/Animal
    Management, Birkhäuser Verlag, Basel, Boston, Stuttgart. (In press).

Huber, A. and Fölsch, D.W., 1978. Akustische Ethogramme von Hühnern. Die
    Auswirkungen verschiedener Haltungssysteme. Tierhaltung, Vol. 5,
    Birkhäuser Verlag, Basel, Boston, Stuttgart.

*
All references with detailed summaries in English.

DISCUSSION

J.P. Signoret *(France)*

   Dr. Fölsch has distinguished between 'mutual disturbances'
and 'threat and aggression'. He shows no great difference
between the three housing systems for 'threat and aggression',
whereas for 'mutual disturbances' it is very clear. Can he
just explain what the difference is?

W. Fölsch *(Switzerland)*

   Threat and aggression is quite clear I think. Mutual
disturbance may be, for example, when animals are in close
contact with each other, some animals are at the feed trough
and another animal wants to come to it. It is not necessarily
threatening, but it is trying to come between them, pushing the
other hens away.

J.P. Signoret

   So there are no agonistic interactions, so it means
disturbances that take place without aggression?

W. Fölsch

   I did not say that; I only said there is a physical
interference.

J.P. Signoret

   Without threat or aggression?

W. Fölsch

   I have not related it; I have only observed the effect
that the animals are pushing each other away. This is almost
entirely related to space; threat and aggression is quite a
different thing.

J. Fris Jensen *(Denmark)*

Can you describe the type of cages you have used in more detail? Also, can you say a little more about stocking density; you seemed to be comparing different group sizes, two, three and four hens per cage. But did each bird have the same number of square cm?

W. Fölsch

We used commercial cages in three tiers. They measured 38 cm in width. It was critical to measure the depth because we had the food trough and the material which prevented the animals coming too near to the front. It was approximately 41 - 43 cm. There was a slope of 14%. The height was about 43 cm in the middle. Obviously it varied somewhat because of the sloping floor. Normally these cages are stocked with 4 hens; we had two, three and four hens per cage.

R. Tauson *(Sweden)*

Were the partitions between the cages solid or wire?

W. Fölsch

They were wire.

R. Tauson

We found quite a variation in effect on feathering according to the kind of partition between the cages. If you use solid divisions you do not get so much wear on the feathers, especially on the wings.

J.P. Signoret

You showed a picture of the legs of one fowl. Have you analysed leg damage as you have done with feather damage?

W. Fölsch

No, that work is not finished yet. We have some qualitative indications as to which kind of damage relates to each system.

L. Spanoghe *(Belgium)*

You showed us examples of bumble foot as a typical health problem with animals in bad wire floor cages. Have you recorded losses from disease and disease problems in the other systems?

W. Fölsch

I wanted to show the qualitative change of the feet in different housing systems. This so-called bumble foot has different origins. It is seen frequently in wire netting floored cages.

L. Spanoghe

You have shown us one typical health problem, in one system. What about contagious and parasitic diseases in all the systems?

W. Fölsch

Of course there are a lot of other factors involved. I cannot cover them all today; I can only bring some examples to show perhaps where we should continue research and collect data.

W. Bessei *(FRG)*

You have shown us a picture of typical feather loss in cages but I think this type of feather loss is more likely to be caused by feather pecking than by physical damage. I do not believe that feather pecking is only a problem in battery cages.

W. Fölsch

It was not my intention to interpret the results as to whether the loss is caused by feather pecking or physical damage but it was quite clear that with a higher stocking density there was increased feather loss.

W. Bessei

In this case I think it was due more to the group size.

W. Fölsch

This may be so. There may be several causes but it was informative for me to show that feather loss is correlated to the stocking density. I think it is important to have certain relationships with stocking densities, to rearing systems and to housing systems.

I.J.H. Duncan *(UK)*

Dr. Fölsch, you say that you do not want to interpret your results but if you take a call and name it the 'friendly call' and then suggest that it is lower in one system than another, you are making an interpretation by inference. Would it not be better to say that you have looked at three calls and there is a difference?

W. Fölsch

I would say that the calls are correlated with a certain situation. When you have aggression with hens then you also have a typical vocal expression. It is the same when you have a friendly situation. That we have quantified, and I would say that the adjectives 'friendly' and 'dominant' express the situation fairly well. Of course there is an element of interpretation there but I would like to stick to that.

I.J.H. Duncan

I consider that to be extremely subjective. It is a different case with post-laying cackle which has been examined in some detail, and the physiological background is known for that particular call. It is very subjective to say that a call is a 'friendly' call because you think that the birds who are making it are in a particular state.

W. Fölsch

If you have 60 people out of 100 making the same subjective assessment, then that is a fairly high percentage and I
would say you are coming away from personal interpretation and
getting more objective information.

B.O. Hughes *(UK)*

I am interested in this 'dominant' call as well. I am
not sure what this call relates to in the English literature.
Is it the short, rather hard call that the birds make as they
approach each other?

W. Fölsch

Yes, that's right.

B.O. Hughes

I thought that was it. I have not done any quantitative
work on whether it occurs before or after threats or pecks, or
when the bird is at a certain distance. I agree with Dr. Duncan
that it is still open to question whether this call is associated with aggression as far as we are concerned, but if one
accepts that it is so, it is interesting that it correlates
with the fact that in cages you get less threatening and less
pecking at increased stocking densities. Your work suggests
that this call increases which makes good sense in that situation and may suggest that there is a transference from overt
aggression to covert aggression. Is that a reasonable interpretation?

W. Fölsch

It may be that birds do utter a particular call before
making a threat, but I would stress that we did not compare
different stocking densities in the battery cages concerning
this aspect.

R. Tauson

In our work at Uppsala we do a lot of scoring of birds
for feet, feather wear and so on. I am very aware of the
problem of showing pictures that are typical for different
systems. I would like to stress three points. First, the
design of the systems, both the cages and the floor systems and
the deep litter systems. There is a great variation between
these. Also, there is the variation between birds in the
system. As Dr. Signoret pointed out, it is very important to
know how the feathering can differ between two birds in the same
system; we must be very aware of that and also of how the
feathering can vary between floor systems. As Dr. Bessei
pointed out, you can easily get feather pecking in all systems
which can also give a variation. Then there is the question
of the behaviour of the birds. Birds vary in their behaviour
in cages, in how they move their feet, and this can account for
whether or not they get foot damage. So, these are three
points: the design of the cage, the variation between birds and
the specific behaviour of the birds in different systems.

W. Fölsch

On the last point, as I told Dr. Signoret, we did not
quantify damage to the feet but the damage to plumage has been
checked not only in our research project but also on egg
producing farms and data on 200 or 300 hens has been published,
together with detailed descriptions of the different housing
systems*.

J. Fris Jensen

You also described a deep litter system. Have you tried
to evaluate the quality of the deep litter? Have you a pit in
your deep litter system or was the whole floor covered in deep
litter?

----

* Volume No. 9, Series 'Tierhaltung'.

W. Fölsch

We have done our research on different farms. On our
experimental farm we had no pit in order to keep the same floor
size for deep litter and wire netting. In the commercial farm
and other sites we had deep litter with pits. We did not
choose a particular quality of deep litter because the quality
changes during the year. We were more concerned to find an
egg producing farm where hens were kept both in batteries and
in deep litter so that the birds had the same climatic environment and the same standard of care.

J. Fris Jensen

On the farms with deep litter pits, what proportion of
the area was taken up by the pit?

W. Fölsch

On one farm, 50%; on another, 75%.

I.J.H. Duncan

I would like to come back to subjectivity. You said that
locomotion was inhibited in inside systems compared to outside,
free range, systems. Would it not be better to say that hens
walk about more outside than inside? If I was a man from Mars
looking at hens inside and outside, I could just as easily make
the interpretation that the poor hens outside are being forced
to walk about a lot more. You are making an interpretation if
you say that by keeping them inside locomotion is being
inhibited. It may be that when they are outside they are moving
about because they are looking for predators, because they are
frightened. Just because many people make the same interpretation that does not necessarily make it correct. When monkeys
in the zoo show their teeth, 99% of the population would say
they are smiling, but you know and I know that in fact they
are threatening.

W. Fölsch

I agree with you fully. That was the reason why I showed
first the percentages of the behavioural activities which is
the basis of the information, but I am sure that it is also
important to consider the environment and how the activity is
performed. For example, whether nest preparing is done by
taking up straw or branches or by pulling at the bird's own body
to take out feathers. We have to describe exactly what is done.
I agree that we have to quantify exactly what we see, and before
we can do that we have to have a qualitative description of
that behaviour.

G. Martin (FRG)

Dr. Duncan, I think that you are overlooking the fact that
many of these behaviour patterns are innate - the hen must do
these things.

I.J.H. Duncan

I think it is true that we must examine the causation and
the expression of these behaviour patterns very carefully, but,
for example, we know that nesting behaviour is usually preceded
by locomotion, but we do not know whether that locomotion would
be equally well expressed by the bird walking in a big wheel,
or whether it has to walk in a straight line or a zig-zag line,
or a particular distance. We do not know whether it would have
the same effect if it walked round and round a cage. There is
some evidence that birds prefer to get away from their flock
mates and perhaps that has something to do with the locomotion.
Perhaps walking round and round a cage would have the same
effect. Similarly with nesting; some birds do show all the
motor patterns in cages; they show nest building movements
although there is no nesting material there. Does the performance of these motor patterns in that situation suffice the bird?
We do not know.

B.O. Hughes

We will be reviewing these problems in the next paper, so perhaps after that would be a better time to discuss them.

R. Tauson

Dr. Fölsch, in these experiments did you do any statistical analyses of the correlation between feathering and production and sore feet?

W. Fölsch

We have the production figures comparing deep litter to battery cages and the evaluation between different densities, four, three and two hens. There is virtually no difference between deep litter and cages, but there is a slight difference between four hens per cage and the other densities.

R. Tauson

I was asking for a correlation between production and feathering within systems; or a correlation between production and feet damage within systems.

W. Fölsch

We have not made such correlations.

R. Tauson

In your opinion, do these foot blisters cause the birds to suffer?

W. Fölsch

When I go into a hen house and see birds continually standing on one foot, of course I must assume that they are not doing so just because they like it. If any animal has an injury on a part of its body it takes care of it. Of course, it may be a personal interpretation, but according to my opinion and to my education as a veterinarian, if an animal is blistered in this way it is hurt. All veterinary work is done by

interpretation of symptoms; in the same way that we can interpret the heart activity of a cow we can interpret the sounds of
hens - we have to do so.

I.J.H. Duncan

Yes, obviously in some situations it is perfectly possible
to interpret the sound a hen is making. However, in other
situations we must be very exact; we must describe exactly the
social environment of the hens, the particular time that the
call is given; we must give, if possible, a sound spectrograph;
we must try to describe the things which affect that call and
then make an interpretation.

W. Fölsch

As you know, we started this work in 1973. We have
observed wild junglefowl as well as Shaver Starcross; we have
tape recorded the calls of chickens from the first day of their
lives, we have recorded broody hens, and so on. We collected
as much information as we could on tape recorders. Afterwards
we got the sonograms and then after about one and a half years
we put a microphone in each box and collected sounds of each
laying hour from each box; for a certain time all calls from
that box were recorded. Then we were able to evaluate and
count how many calls were uttered by the hens. I cannot explain
the whole of the technique now, but we have published it, as
you know. Nowadays when we go into a unit we have the experience to be able to differentiate the calls.

I.J.H. Duncan

I am sorry to keep on about this, but I think there is a
difference between describing a call as a 'post-laying cackle' -
purely objectively - and describing a call as a 'friendly' call.
I think that is a subjective interpretation.

W. Fölsch

Therefore we have the sonograph to characterise the call.
At the same time we have a description of the situation, when

the call was uttered. From all these observations we are able
to establish the kind of calls which are typical in various
situations.

I.J.H. Duncan

I accept that; I accept that many people have done work
on calls. The point on which I am arguing is the name you give
to these calls. For example, day old chicks commonly give two
types of call. Originally these were called 'distress' calls
and 'pleasure twitters'. It is much more objective to call
them 'peeping' and 'twittering', which is just an objective
description; it does not make any interpretation about the
underlying motivation.

W. Fölsch

These names were not given by us; they were given by
Baeumer many years ago.

I.J.H. Duncan

And that was the mistake that Baeumer made. Baeumer did
a good job in describing the call, but his mistake was to name
it a 'friendly' call.

W. Fölsch

But he also tried to fix it on a tape recorder which was
difficult at that time.

I.J.H. Duncan

Yes, and that was excellent.

W. Fölsch

But we have now done it; we have the subjective name –
you can call it the 'q' call if you prefer – but we also have
the sonogram and we can show at which time of the day a given
call was uttered and under which circumstances.

I.J.H. Duncan

That is fine; I would much rather call it the 'q' call.

W. Fölsch

To me, 'q' call means nothing; I think it is better to have a descriptive name for typical calls in typical situations.

J.A. Hill *(UK)*

May I ask whether you used the same strain of bird in all the experiments you have described?

W. Fölsch

We started with so-called wild junglefowl, free range, and Shaver Starcross. In the second experiment we had LSL normal selected Leghorns. On the egg farms there were normal selected Leghorns as well as Shaver Starcross.

J.A. Hill

Thank you. I would like now to come back to the question of interpretation. Do you consider that simply pursuing the idea of quantifying different behaviours in different systems is going to help us in the long term, to any extent? I think we would all expect to see a different amount of any particular behaviour pattern in the various systems. When we come to interpret this difference, is there any evidence that the fact that a bird is not showing as much of a particular type of behaviour in one system, it is necessarily a welfare disadvantage to that bird? I think this is an important question.

W. Fölsch

It is very important. We have realised this since the beginning of our work in 1973. Hens do not dust bathe in batteries, but is it simply because there is no reason to? This is why we have had to make detailed observations of the birds from when the lights go on until the last hour.

J.A. Hill

But how do you interpret whether or not it is disadvant-
ageous to the bird not to be showing a particular behaviour in
a given environment?

W. Fölsch

When you have sand bathing on a wire netting floor, then
you have the information that the hen will do it even though
the environment is not adequate for the behaviour, and it will
be damaged in that environment. Again, if the hen runs around
on material which is not adequate for the feet they get hurt
and blistered feet. I would say it is possible to make an
interpretation from this.

K. Vestergaard *(Denmark)*

I would like to return to the question of feather damage.
You showed results from different systems. It seems that going
from one system to another, many environmental and social factors
are changed. I agree that first we had to look at the various
systems and see where the differences lay. But now I think we
should move on to the next stage and look at different factors
in the environment. How does flock size affect feather pecking?
How does density affect it, and the quality of the floor, for
example? You probably know that in Denmark we have studied the
effect of flock size and the effect of wire versus litter floor.
We found that a wire floor increased the damage very much. We
also found that increased density affected the amount of damage,
especially on the wire floors. So this might be the next step,
probably followed by establishing exactly how the damage occurs,
by pecking, wear and tear, or whatever.

W. Fölsch

Yes, but we cannot ignore the practical problems. For
example, many existing cages will be on farms for the next ten
years and we should be able to say whether four, or three, or
two hens should be put in them. We have different ways of
approaching the problems - and this is good, we need that.

C.C. Brantas   *(The Netherlands)*

I agree with Dr. Duncan that it is very important to be
objective because of the danger of subjectivity. But there is
also a danger of over-objectivity. In his published work
Dr. Duncan refers to threats as 'type 1' and 'type 2' - surely
this is also an interpretation. I suppose, Dr. Duncan, when
you have an encounter with a dog you do know whether that dog
is friendly towards you. If not, be careful! This is not a
new problem. Research workers have been wary of making subjective assessments for many years, but if they become over-
objective then the work becomes valueless.

M. Zanforlin   *(Italy)*

May I intervene instead of Dr. Duncan please?

I.J.H. Duncan

Thank you very much!

M. Zanforlin

I would agree with my colleagues that to use the term
'friendly' call and so on is a very short way of describing
behaviour, but we do not always realise the emotional implications that this kind of terminology may give rise to in the
people who are not actually there in the particular situation.
I recall work designed to demonstrate the presence of consciousness in three day old chicks. The definition of consciousness
was if they could change their behaviour in response to an
outside stimulus, then there is consciousness. So learning
became synonymous with consciousness. If you go down the zoo-
logical scale to the *planaria*, the *planaria* is able to learn
and so the *planaria* has consciousness! That is the danger.
I agree that between research workers it is convenient to use
the term 'friendly' call but when this word is heard by the
general public then they only consider the emotional implications
of the term, rather than the scientific definition which we
understand among ourselves.

H.C. Adler

We must close this session now and I thank Dr. Fölsch for his very interesting presentation which produced a very lively discussion.

THE ASSESSMENT OF BEHAVIOURAL NEEDS

B.O. Hughes

ABSTRACT

In this short paper the possibility of being able to define the behavioural needs of the laying hen is discussed. A unitary model of motivation is described, and it is concluded that each behaviour pattern must be considered as a separate entity. The relative roles of internal and external factors must be gauged, and a case can be argued for giving hens the opportunity to express those behaviours which are governed largely by internal factors.

R. Moss (ed.), The Laying Hen and its Environment, 149-166
Copyright © 1980 ECSC, EEC, EAEC, Brussels-Luxembourg. All rights reserved.

One of the objects of this conference is to determine
whether or not there is a consensus among poultry ethologists
concerning the behavioural needs of the laying hen. A secondary
question is whether the battery cage can meet these requirements. Hughes (1976) has argued that one way of assessing
welfare is to compare the ethogram of the battery hen with that
of a hen in an environment which is assumed to be ideal. Such
an environment is not readily available, and indeed may not
even exist, but in behavioural terms a good deep litter system
may approximate to it, and has been used as a reference point
for a number of studies (Bareham, 1972; Black and Hughes, 1974;
Hughes and Black, 1974; Eskeland, 1977; Fölsch, 1977). These
studies concur in demonstrating that there are marked differences in the ethograms of birds housed on litter or in cages,
mainly in quantitative terms, though there are some qualitative
differences too. However, although the facts are not a major
source of contention, not all ethologists would interpret them
in the same way. There is a dichotomy between those who, like
Thorpe (1965) or Martin (1977), consider that the prevention
of all innate behaviour patterns is undesirable, and those who
argue that the ontogeny of the behaviour must be understood
before its importance can be assessed. This view was first put
forward by Wood-Gush (1973) and has been reiterated by Wood-
Gush et al. (1975). They suggest that many behaviour patterns
are partly dependent on internal factors which may periodically
vary in strength but which do not accumulate with time, and
partly dependent on external stimuli which act as releasers and
result in overt behavioural expression. Some behaviours can
occur in very barren surroundings in the absence of an obvious
releaser; for example, dust bathing in battery cages, though its
incidence is much lower than it is on litter.

Now, these findings can be interpreted in two ways.
Taking a Lorenzian stand; action-specific energy builds up until
it has to be discharged, even in the absence of a suitable
releaser, and it then appears as a vacuum activity. Using the
mixed-motivation model (Deutsch, 1960; Hinde, 1969) the tendency to dust bathe increases from time to time, but in the

absence of a suitable substrate it generally remains latent,
becoming overt only on the rare occasions when a combination of
stimuli from the environment coincide, and the effective
threshold is reached. In the context of this model, failure to
perform a particular behaviour pattern such as dust bathing
would not result in frustration.

I wish to suggest that the dichotomy, as I have presented
it here, is a false one. Instead, let us suppose that the
Lorenzian model is one extreme, and the mixed motivation model
is the other extreme of a continuum. Each different behaviour
pattern of the fowl will lie at a particular point on the
continuum. At one end the behaviour will be triggered largely
by external events and hardly at all by internal variables. In
the centre internal and external events will be of equal importance, while at the other end internal tendencies will largely
govern the animal's behaviour, with little influence being
exerted by external stimuli. We now have a unitary model, and
the Lorenzian approach can be regarded as a special case, lying
at one end of the continuum. If this unitary model is accepted
there are several implications.

Firstly, one cannot generalise. Each behaviour pattern
must be considered as a separate entity, and an attempt made to
determine to what extent it is controlled by internal and
external factors. This may be difficult in practice because of
the close interaction between them.

Secondly, it may nevertheless be possible to divide the
continuum at certain points, and to recognise various categories
of behaviour. For the sake of argument, a three-way classification could be used where:

1. External stimuli are of over-riding importance.
2. Internal and external variables interact closely
    and are both important.

3.   Internal factors are most influential.

Thirdly, the internal factors may also compete among
themselves. Many behaviour patterns do not occur at a particular time because they are being inhibited by others of higher
priority. For example, a hungry bird will normally not preen,
but spend its time searching for and ingesting food. When its
hunger is satisfied it may preen, not because there has been
any change in the internal or external factors directly
affecting preening, but because hunger is no longer inhibiting
preening and the hen now pays attention to the preening
factors. One may be able to explain some of the differences in
the incidence of behaviour patterns in different environments
in such terms. However, the observed facts do not always fit
neatly into such a framework. One would predict that in
environments where food was freely available and could be
rapidly ingested, hens would be engaged longer in activity
which was of a low priority, whereas in fact such is often not
the case.

Fourthly, human involvement can manipulate the expression
of these behaviour patterns in two ways. Environmental change
can alter the influence of the external stimuli, while genetic
selection can affect the internal tendencies.

I should now like to consider a number of behaviour
patterns in the context of this model, beginning with the first
category.

Escape behaviour is a direct response to unpleasant or
frightening stimuli, and therefore can be regarded as being
mediated largely by external factors, though even in this case
the response can be modulated by the internal state of the hen
(see for example environmental effects on fear (Hughes and
Black, 1974; Jones and Faure, in preparation)). Agonistic
behaviour is stimulated by the presence or the activity of
another bird, but again, the nature and extent of its expression
will be influenced, to an extent, by internal factors.

When we come to the second category, the sexual crouch is a good example of a behaviour which depends on the close interaction of internal and external factors, for it is observed only when·the hen is in an appropriate hormonal state and when she is presented with an adequate stimulus, such as an adult male. It can also be triggered by less suitable stimuli, such as human beings, but usually only in newly-mature hens. This suggests that there may be an additive effect, and that sub-threshold stimuli can be effective if the internal motivational factors are particularly strong. If similar reasoning is applied to dust bathing, its appearance in cages under conditions when external stimulation is minimal suggests that internal factors must be especially influential at times. In the same way the increase in 'playing with food' and in pecking at the feathers of other birds or at the cage may represent the expression of a pecking tendency. This would be directed at vegetation if the hen was on range, or at litter if she was housed inside on the floor, whereas in cages it is released by what would normally be regarded as inappropriate stimuli.

I shall go on from this to suggest that if a behaviour pattern of this second type is observed in a barren environment, that is, where the external stimuli which normally release it are very weak, then it is probable that the internal stimuli are especially strong, which may be taken as *prima facie* evidence of behavioural need. It must be emphasised that this is in no way evidence that the hen is in distress. Indeed, it can be argued that because of the very fact that the behaviour can be carried out, her behavioural needs are being satisfied and frustration is not occurring. To argue that frustration is present, other evidence is necessary, and in this context it is suggestive, no more, that some types of feather pecking observed in caged hens are remarkably constant in nature, which may have led Wood-Gush and Rowland (1973) to speculate that they have analogies with the kind of stereotype induced by barren surroundings.

When behaviour occurs in an inappropriate context, or in a wholly fixed fashion such that it is unrelated to its environmental context, this suggests that the normal external stimuli are not influencing it and it is a reasonable assumption that it is driven largely by internal factors. Nest building in hens about to lay (Wood-Gush and Gilbert, 1969), where imaginary material is placed on the hen's back or carefully re-arranged around her, is such an example. Another example is the need to sleep, which can occur in the absence of the external stimuli that normally precede it. The occurrence of a stereotyped response suggests the frustration of a behavioural need. Sometimes the causal chain is clear: pacing in light-hybrid strains in cages indicates their inability to find an appropriate nest site (Duncan, 1970), while stereotyped air-pecking and cage-pecking has been observed in calcium-deprived birds (Hughes, 1973). Sometimes the causal chain is less clear: Eskeland (1977) reported that cage-pecking was frequent in caged hens, but commented that it was unclear whether the reason was frustration or a reduction in environmental stimulation.

I would like to put forward for discussion the thesis that it is _desirable_ that provision be made for some of the behaviour patterns classified in Figure 1 under category 2 to be carried out, perhaps including dust bathing, but excluding sexual crouching, and _essential_ in the case of category 3. The criteria for including a behaviour pattern under category 3 would be that: either

1. There is clear evidence of frustration.

2. The behaviour emerges in an abnormal or distorted form. This depends on our ability to define the limits of normal behaviour, and more work is required on this.

Further evidence may emerge from studies of hens' preferences for different environments, which are potentially a powerful tool for determining behavioural needs, but require cautious interpretation. Research carried out by Hughes (Hughes, 1975; 1976; 1977; Hughes and Black, 1973) and Dawkins (1976; 1977; 1978) has shown that hens choose:

a. Floors which provide more points of support for their feet.

b. Large as opposed to small spaces.

c. An outside run rather than a battery cage.

d. Familiar neighbours rather than an empty cage, but an empty cage over strange hens.

e. A litter floor rather than wire mesh when the choice is irreversible.

Interpretation must be cautious because these preferences are relative rather than absolute, are strongly influenced by the previous experience of the bird, and are affected by testing method. For example, litter is selected when the hen makes her choice and then has to remain in the chosen environment for an 8 h period, but the preference was much weaker when the hen was able to move from the wire floor to litter and back again at will. This situation provides an interesting example of the interaction between internal factors and external stimuli, because even wire-reared hens, who otherwise spent most of their time on the mesh floor, laid almost all their eggs in litter. This suggests that access to litter during the pre-laying period may be an important requirement for the hen.

156

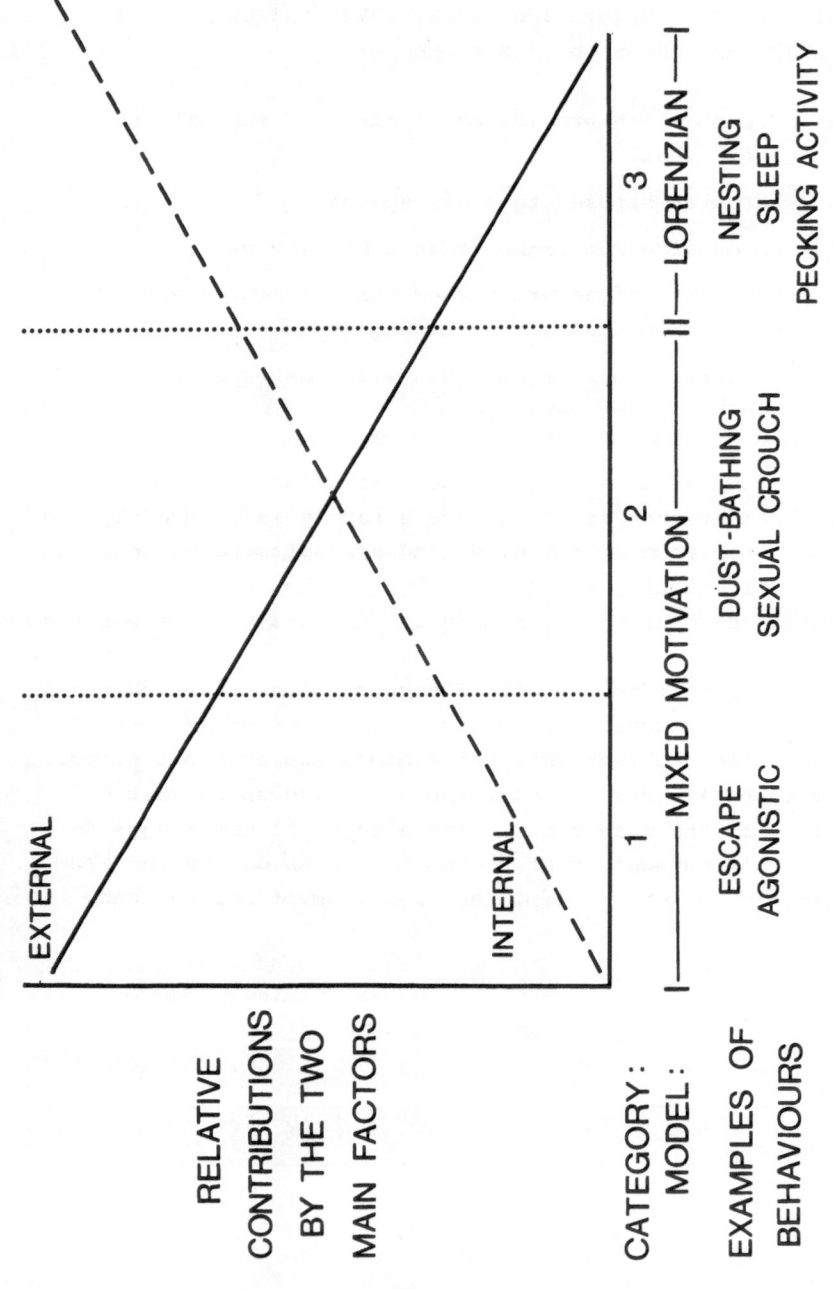

Fig. 1   The unitary model

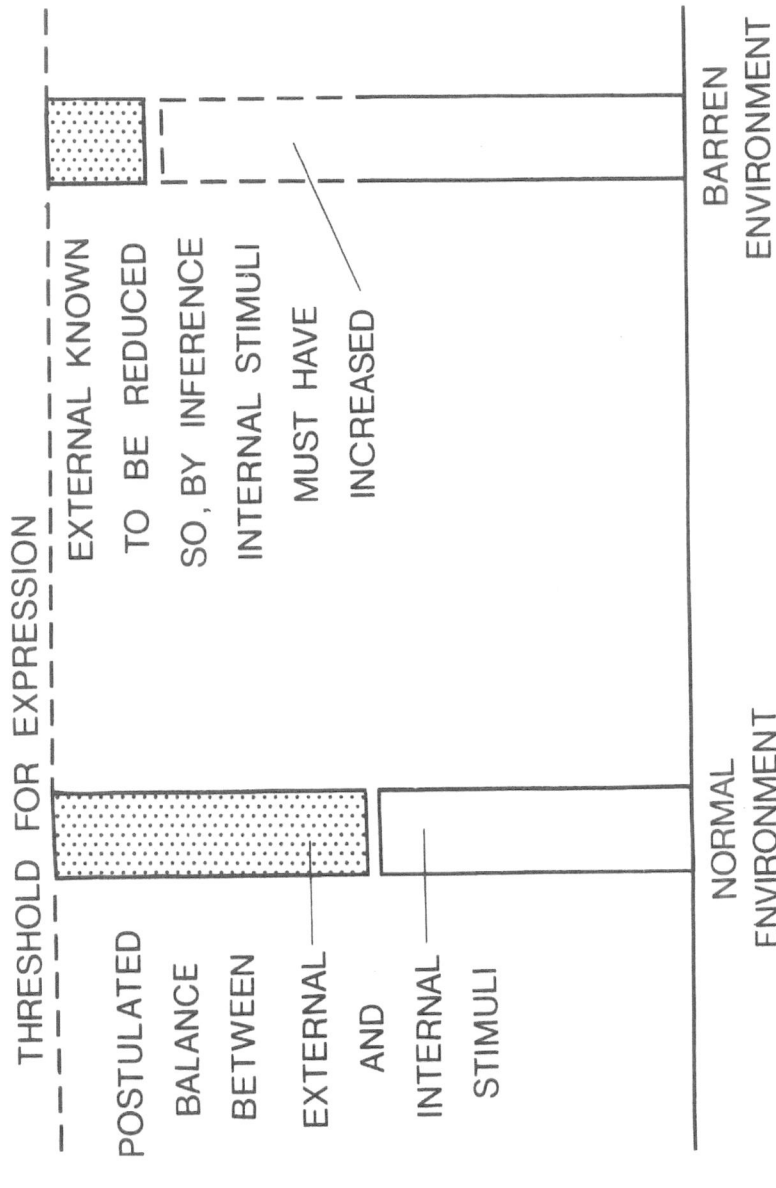

Fig. 2   Schematic diagram showing how internal stimuli have to increase before behaviour can be expressed in a barren environment.

158

REFERENCES

Bareham, J.R., 1972. Effects of cages and semi-intensive deep litter pens
on the behaviour, adrenal response and production in two strains of
laying hens. Br. Vet. J., 128: 153-163.

Black, A.J. and Hughes, B.O., 1974. Patterns of comfort behaviour and
activity in domestic fowl: a comparison between cages and pens.
Br. Vet. J., 130: 23-33.

Dawkins, M., 1976. Towards an objective method of assessing welfare in
domestic fowl. Appl. Anim. Ethol., 2: 245-254.

Dawkins, M., 1977. Do hens suffer in battery cages? Environmental
preferences and welfare. Anim. Behav. 25: 1034-1046.

Dawkins, M., 1978. Welfare and the structure of a battery cage: size and
cage floor preferences in domestic hens. Br. Vet. J., 134: 469-475.

Deutsch, J.A., 1960. The Structural Basis of Behaviour. Cambridge:
Cambridge University Press.

Duncan, I.J.H., 1970. Frustration in the fowl. In: Aspects of Poultry
Behaviour. Ed. by B.M. Freeman and R.F. Gordon, Edinburgh.
British Poultry Science.

Eskeland, B., 1977. Behaviour as an indicator of welfare in hens under
different systems of management, population density, social status
and by beak trimming. Meldinger fra Norges Landbrukshogskole.
Vol. 56, Nr. 7.

Fölsch, D.W., Niederer, Ch., Burckhardt, Ch., and Zimmermann, R., 1977.
Untersuchungen von Legehennenhybriden unterschiedlicher Aufzucht in
verschiedenen Haltungssystem während einer Legeperiod von 14 Monaten:
Wirtschaftlich relevante Aspekte. Tierhaltung. B and I. Volume 1.

Hinde, R.A., 1969. The bases of aggression in animals. J. Psychosom.
Res., 13: 213-219.

Hughes, B.O., 1973. An increase of activity in domestic fowls produced
by nutritional deficiency. Anim. Behav., 21: 10-17.

Hughes, B.O., 1975. Spatial preference in the domestic hen. Br. Vet. J.
131: 560-564.

Hughes, B.O., 1976a. Preference decisions of domestic hens for wire or
litter floors. Appl. Anim. Ethol., 2: 155-165.

Hughes, B.O., 1976b. Behaviour as an index of welfare. Proc. V. Europ.
Poult. Conf. Malta, 11: 1005-1018.

Hughes, B.O., 1977. Selection of group sizes by individual laying hens. Br. Poult. Sci., 18: 9-18.

Hughes, B.O. and Black. A.J., 1973. The preference of domestic hens for different types of battery cage floor. Br. Poult. Sci., 14: 615-619.

Hughes, B.O. and Black, A.J., 1974. The effect of environmental factors on activity, selected behaviour patterns and 'fear' of fowls in cages and pens. Br. Poult. Sci., 15: 375-380.

Martin, G., 1978. Ethologie und Ethik und ihre Konsequenzen für die moderne Nutztierhaltung. In: The Ethology and Ethics of Farm Animal Production. Ed. by D.W. Fölsch, Basel and Stuttgart, Birkhauser Verlag.

Thorpe, W.H., 1965. The Assessment of Pain and Distress in Animals. In: Report of the Brambell Committee. Cmnd 2836. London, Her Majesty's Stationery Office.

Wood-Gush, D.G.M., 1973. Animal Welfare in modern agriculture. Br. Vet. J., 129: 167-173.

Wood-Gush, D.G.M. and Gilbert, A.B., 1969. Observations on the laying behaviour of hens in battery cages. Br. Poult. Sci., 10: 29-36.

Wood-Gush, D.G.M. and Rowland, D.G., 1973. Allopreening in the domestic fowl. Rev. Comp. Animal, 7: 83-91.

Wood-Gush, D.G.M., Duncan, I.J.H. and Fraser, D., 1975. Social stress and welfare problems. In: The Behaviour of Domestic Animals. Ed. by E.S.E. Hafez. London, Bailliere Tindall.

DISCUSSION

K. Vestergaard (Denmark)

May I ask why you put dust bathing into your category 2?
One of your arguments for putting patterns into category 3 was
that behaviour should be distorted or abnormal. Dust bathing
in cages has not been studied very closely but from the evidence
we have it would appear to be 'distorted'.

B.O. Hughes (UK)

Yes, that is a good point. There are two ways of looking
at it, either in qualitative or quantitative terms. Quantitatively, the level of dust bathing in cages is much lower than
it is on litter. From my work I would have said it is less
than one tenth. So it is obvious that external stimuli play an
important role in triggering it off. Equally, it does occur in
cages, so therefore there must be internal factors which can
trigger it off. I did not consider it distorted and abnormal
in the same way as nesting behaviour, but that is a qualitative
point. As you say, that argument would help to push it over
the boundary line into category 3.

K. Vestergaard

It seems to me to be important to distinguish between
motivating stimuli and eliciting stimuli. Internal factors are
very significant as regards motivating stimuli. In the
environment you can have external stimuli which are both
eliciting and motivating.

B.O. Hughes

I think this is probably more a question of definition.
I just wanted to keep it as simple as possible.

W. Fölsch (Switzerland)

I think it is important to define normal behaviour,
otherwise we cannot say what is abnormal. Normal behaviour

patterns are shown in an environment which is equivalent to the original ecology of the animal. As far as abnormal behaviour is concerned, I think we can make a comparison between the percentage of various behaviour patterns in relation to the percentage in a biological environment. We can also look at the sequence of behaviour, for example, with dust bathing; whether it is begun several times a day or begun, run through, and ended after perhaps twenty minutes.

B.O. Hughes

May I deal with your second point first. It is a very important point. You are right, it does yield a lot of additional information if one looks at the dust bathing sequence of a bird on litter compared to a bird in a cage. If the one in the cage is quantitatively very different, it stops short after a very brief time and does not go through the normal pattern, that does give us additional information as to how to classify it.

I do not agree with your first point. You are arguing that the hen should be put back in its original ecological niche. It is valid for junglefowl but I do not think it is true for the modern laying hybrids. There is no such thing as a 'normal' environment for them because their genetic quality has been changed over the last hundred years; it is still constantly changing, and there is no balance between the genetic make-up of the birds and the environment in which it is being kept. In another 50 years the normal environment of the laying hen could be the battery cage, if we continue selecting along these lines.

I was careful to say that a good deep litter system might approximate to a normal environment, because it incorporates most of the elements for which the bird was selected up until the last ten years. I believe I am right in saying that as regards the laying hen, it is only in the last five to ten years that the poultry geneticists have been selecting birds intensively in battery cages.

Another point is that our modern hybrids originate from a very varied geographical and climatic background and so it is not possible to lay down a 'normal' environment.

W. Bessei *(FRG)*

I think it would be preferable to talk about adaptive or non-adaptive behaviour rather than normal and abnormal. I think we should study the question of whether or not a bird is able to adapt to an environment.

B.O. Hughes

I do not think there is any basic disagreement between us; we are probably saying the same thing in different terms.

I.J.H. Duncan *(UK)*

In relation to adaptive and non-adaptive, I think an example would be in nesting behaviour, where some hens in cages show 'normal' nesting behaviour if you consider that the motor patterns are the same, the sequence of the motor patterns are the same, and they appear to go on for the same length of time. One thing follows another and leads to oviposition. Other hens in cages do not show this pattern. The first part of nesting behaviour, the increased locomotion, is continued; it develops into stereotype pacing which we know from other experiments is symptomatic of frustration, and we have some evidence that oviposition is delayed. Would you agree that in that case the first type of nesting behaviour in cages is 'normal' and the second type 'abnormal'? The first type is adaptive and the second type is not adaptive.

B.O. Hughes

I would agree with all that; I think that is a very clear way of putting it. I do not know whether it gets us any further in distinguishing between abnormal and non-adaptive. I suspect the two terms are very nearly synonymous. It may not be very helpful to try to distinguish between them.

I.J.H. Duncan

I think there is a difference because there are some
people here who would not regard the first type of nesting in
cages as normal.

B.O. Hughes

Possibly that is true. You are speaking here of the kind
of behaviour one sees in medium hybrids, which settle down,
ruffle their feathers and appear not to be agitated, but nevertheless do not go through all the complex sequence of movement
that they would do were they on litter?

I.J.H. Duncan

They go through the normal sequence but it is performed
as a vacuum activity. There is no nest building material there.

B.O. Hughes

It is not completely normal, it cannot be.

I.J.H. Duncan

Well is it, or isn't it?

B.O. Hughes

It is quantitatively different - the actual length of the
sequences is almost certainly different. In fact, I don't
think a great deal has been done on the precise sequential
pattern of the behaviour in the same bird in litter and on
wire. There are practical difficulties here. Wood-Gush has
done an experiment in which the same bird was placed in a
battery cage, one day with a wire floor, the next day with a
litter floor, and so on. So the eggs were laid alternately on
wire and litter floors. He found that in the case of light
hybrids, the litter floor had an important impact on the amount
of pacing. But certainly, the light hybrid did not behave
'normally' even on the litter floor. The medium hybrid also
showed changes in behaviour with the change of floor, which

would suggest that its behaviour on the wire floor was not totally normal, that altering the external stimulus did have some effect on the pattern of its behaviour. So, I cannot accept that the brown birds showed a precisely normal sequence on wire in that experiment.

R. Moss (UK)

We are looking for the ethological and physiological needs of the hen. Listening to the discussion, it seems to me that we are possibly looking for Dr. Signoret's 'ecological niche' in three situations. Perhaps we ought to be thinking about the 'normal' in those three situations: one on free range, one on deep litter and one in cages. In looking at that, the normal is clinical health without frustrative or distorted behaviour. Would that be a basis on which you could work?

B.O. Hughes

Well, I was talking about distorted behaviour, but perhaps we should be talking about non-adaptive rather than distorted. This may help to get round the difficulty of defining 'normal' and 'abnormal', and therefore 'distorted'. Basically, I agree with what you say, Mr. Moss, but I would want to think about it for some time before making a final judgement.

K. Vestergaard

May I refer back to your category 2? We were talking about dust bathing, and you said that in your experience dust bathing in cages was one tenth of that in deep litter. It may not be entirely due to the significance of external stimulation; it may be due to inhibition factors from the other birds in the cages.

B.O. Hughes

I have no wish to be dogmatic about this. It may well be that the external factors have a different degree of influence

in different environments. It may be that on deep litter the
important external motivation is the presence of the dust, but
when you go on to a wire floor, large flock system, some other
factor becomes of importance, and then when you go to cages the
external stimuli are fewer still and it is different again. So
one has to look not only at behaviour patterns in isolation,
but also one has to study them in the different systems before
it is possible to make a sensible interpretation of what is
going on.

M. Zanforlin *(Italy)*

Coming back to the question of a 'normal' environment.
It could be argued that an environment in which the percentage
of mortality was the same as in the wild would be a normal
environment. But the percentage of mortality in cages is much
less than in the wild, so this would be an absurd criterion,
but it is just to show that it is impossible to speak of a
'normal' environment unless we define all the possible variations of the different strains. Each strain, and even each
individual animal, has specific requirements. If we are talking
about the environment of a very large number of animals it is
impossible to define what is 'normal'. It is perfectly evident
to me; I do not know why it is not just as evident to everybody
else.

B.O. Hughes

It is true that you have got variation within a population
and certain individuals are better adapted to a particular
ecological niche than others; also, the ecological niche itself
is not constant, it varies from place to place and time to time.
So, as you say, it is not very meaningful to talk about a rigid
'normal' environment. I agree with you.

I.J.H. Duncan

There is another point we should take into consideration.
Apart from an environment allowing the normal motor patterns to
take place, we must consider what the consequence of these

motor patterns might be. For example, even though an environment allows the normal dust bathing motor patterns to take
place, this dust bathing may then result in injury to the animal,
by feather damage or by the animal breaking its claws on the
wire floor. To take another example, even though a cage
environment allows a certain amount of pecking, that pecking
is directed towards other birds and results in damage to them.

B.O. Hughes

I agree, and this is another danger of the preference
approach whereby allowing the animal to show a clear preference
may not be in its own long term interest.

H.C. Adler

I must conclude the discussion now and thank you all for
your contributions.

ESSENTIAL BEHAVIOURAL NEEDS: THE MIXED MOTIVATION APPROACH

W. Bessei

Although in recent decades it has been generally accepted
that the motivation of behaviour is determined both by genetic
and environmental factors, there is still a tendency to
emphasise the one or the other component. This may be due to
the fact that very little work has been done in the field of
behavioural genetics in farm animals. There is no doubt that
the statistical methods used in animal and plant breeding can
be applied to ethological science as well, as has been shown in
a wide range of studies in laboratory animals such as droso-
phila, rats and mice. The lack of corresponding studies in
farm animals may be either because the number of experimental
animals is often limited and insufficient for genetic analyses,
or because detailed observation methods do not allow large
numbers of individuals to be observed. It may be argued that
genetic analyses of behaviour would lead to the neglect of the
environmental effects. But, as the environmental component is
an important prerequisite in quantitative genetics, the danger
of overlooking the environmental effect is not high.

The genetic analysis of behaviour is not only of academic
value, but will help to answer the question of inherited
behaviour, which is an important factor in the discussion about
animal welfare under intensive husbandry.

I would like to demonstrate this with the example of the
locomotor activity of the domestic chicken.

As to the caging of laying hens, it is argued that the
domestic chicken has a genetically determined need for locomotion,
which has become established during evolution. This need for
locomotor activity cannot be satisfied under restricted
conditions such as cages. Adaptation to these conditions would

R. Moss (ed.), The Laying Hen and its Environment, 167-180

not be possible because of the genetic fixation of the trait, thus, caging chickens would lead to permanent frustration.

Furthermore, the law for the protection of animals in the FRG states that the species-specific need for movement should not be permanently restricted in such a way that the animal is exposed to pain, injury or suffering. In accordance with the above argument, caging of laying hens would conflict with the law for the protection of animals. The argument includes two assumptions which have to be examined more closely:

1. That animals have an innate need for locomotor activity (e.g. that the chick has to perform a genetically-determined quantity of activity).

2. That the amount of locomotor activity is a species-specific trait for *Gallus gallus domesticus* .

Examination of the motivation for locomotor activity shows that locomotion (as a quantitative trait) can be determined by 3 basic components (Figure 1).

1. A genetic component, which is characterised as spontaneous activity.

2. An environmental component, acting through the environmentally determined excitability.

3. The genotype-environment interaction, which exerts its influence via a genetically-determined excitability.

The spontaneous activity can be explained as activity which appears without any environmental influence or stimulation. Its existence has been demonstrated in physiological experiments (Hamburger, 1963; von Holst, 1969) and can be considered as a motivational factor in ethological studies as well. But it must be noted, it is impossible to measure spontaneous activity as such in ethological experiments because it is impossible to avoid entirely any environmental stimulation.

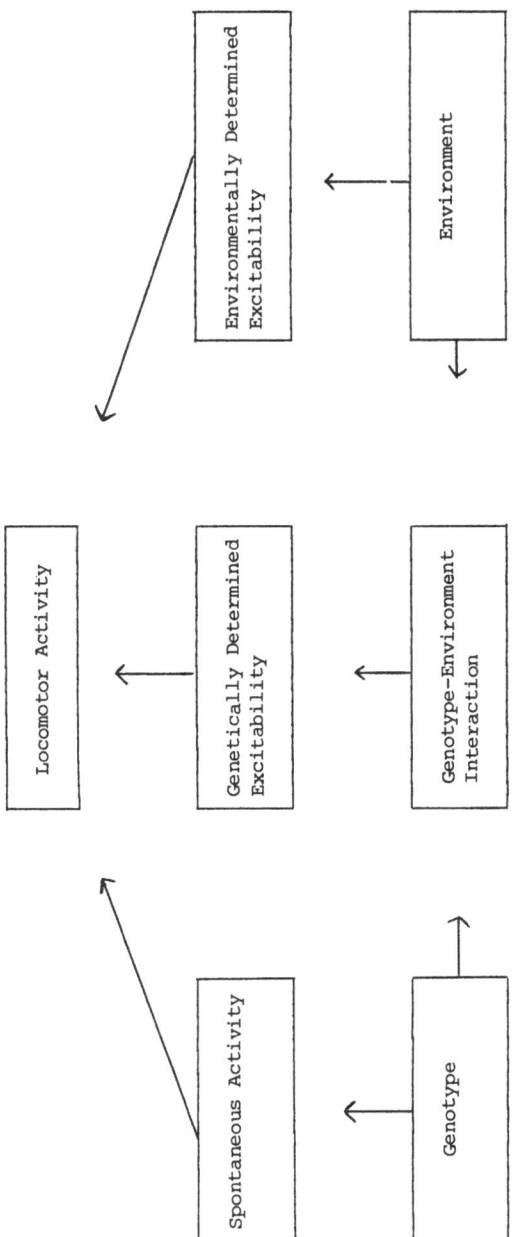

Fig. 1. Determination of locomotor activity in the mixed model

This is also true for the pure environmental component of behaviour. The environmental effect can be shown under extreme rearing conditions or in a conditioning situation; however, it cannot be excluded that the response is partly due to genetic predisposition and there is therefore a contribution from the genotype-environment interaction. Thus, the genotype-environmental interaction will be involved in the expression of the phenotypic behaviour in most cases.

As mentioned above, measuring spontaneous activity and pure environmental activity is not possible in ethological experiments. But the obvious differences in locomotor activity between different breeds, for example, White Leghorn layers and heavy meat types, and even between different lines of White Leghorns (as demonstrated by Jezierski and Bessei, 1978), which can be observed in different environments, may be taken as evidence for spontaneous activity.

The totality of the genetically-determined effects (which includes in addition to spontaneous activity an unknown part of genotype-environment interaction), can be estimated by special biometrical methods, such as calculating the heritability coefficient. So-called innate behaviour can be represented by the coefficient of broad heritability (h_w^2), which is the ratio of the total genetic variance to the phenotypic variance. Calculations of h_w^2 of locomotor activity have been executed in mice (McClearn, 1961). Corresponding investigations in the chicken are not known, but Faure and Folmer (1975) estimated the narrow[*] heritability (h_e^2) of open-field activity in Cornish chicks and found a value of $h_e^2 = 0.15$. The realised heritability over 3 generations was $h_e^2 = 0.35$. In our own investigations into the locomotor activity of adult laying hens (White Leghorn pure lines) the estimated heritability was $h_e^2 = 0.18$. Besides the estimations of heritability coefficients in the domestic chicken, the values in other gallinaceous birds

---

[*]The narrow heritability is the ratio of the additive genetic variance to the phenotypic variance.

may be interesting. Boyer et al. (1973) found in pheasant chicks heritability coefficients of $h_e^2 = 0.51$ to 0.97 for open-field activity. In experiments with adult Japanese Quail the estimated heritability of locomotor activity was $h_e^2 = 0.35$ (Saleh and Bessei, 1978). Thus it can be hypothesised that genetic fixation of locomotor activity would be lower in the domestic chick than in pheasants.

There is more information available about the effect of environmental factors on locomotor activity in the domestic chick. In comparative studies of floor- and cage-housed laying hens, it has been demonstrated that the birds on the floor were more active than those in cages (Hughes and Black, 1974; Bareham, 1972; Eskeland, 1976). In our own experiment however, in which floor and cage birds (birds which had been reared in deep litter or in cages from one day old) were tested in activity cages, the cage birds showed a significantly higher activity than floor birds (Table 1). The floor hens had been given a 2-months adaptation time to single cages in this trial, but in later experiments, where the floor hens were tested after only 12 hours adaptation time, the difference between cage and floor hens was still significant. Black and Hughes (1974) found that the activity of floor hens, which was originally higher than the activity of caged hens, remained at a relatively high level when the birds were transferred from floor to cages.

TABLE 1

THE LOCOMOTOR ACTIVITY OF 2 LINES OF WHITE LEGHORN HENS AS RELATED TO THE REARING CONDITION (JEZIERSKI AND BESSEI, 1978).

Line	Floor	Cage	$\bar{X}$
8	1197	1659	1416
9	1023	1491	1148
$\bar{X}$	1114	1578	1335

Jones (1978) found that male chicks were more active in the open-field than in their home cage.

In a recent experiment we found significantly increased locomotor activity of isolated chicks in the open-field as opposed to birds in groups of three (Table 2).

TABLE 2

THE OPEN-FIELD ACTIVITY OF 2-DAY OLD CHICKS (WL-HYBRIDS) IN RELATION TO SEX AND SOCIAL ENVIRONMENT (1 AND 3 CHICKS PER GROUP)

Sex	Chicks per group		
	1	3	$\bar{x}$
♂	19.3	5.7	12.5
♀	13.7	3.2	8.5
$\bar{x}$	16.5	4.5	10.5

Savory et al. (1978) studied feeding behaviour in free-living Bantam hens and found that locomotor activity was closely related to feeding behaviour. The birds were more active in spring, when they had to catch insects, than in the late summer, when they could find enough food in an oat field.

The effect of genotype-environment interaction has been demonstrated in an experiment with the above-mentioned Cornish strains, which were selected for high and low activity in an open-field (Faure and Folmer, 1975). Table 3 shows the means of open-field activity of 2-day-old chicks as related to the social environment. The difference between the lines was much higher in isolated birds than in groups of three. This supports the assumption that the genetic differences in activity between the two lines is due rather to changes in genetically-determined excitability than to changes in spontaneous activity. If we return now to the previously stated argument of the animal welfarists, we arrive at the following conclusions:

1. Taking into account the considerable differences in
   locomotor activity between the different breeds of
   the domestic chicken, it seems to be impossible to
   define a species-specific need for this trait. It
   would be more realistic to define the need for
   locomotor activity for a given breed in a given
   environment.

2. The genetic component seems to exert its effect more
   on genetic excitability than on spontaneous activity.
   This system enables the birds to adapt their locomotor activity to a wide range of environmental
   conditions.

TABLE 3

THE OPEN-FIELD ACTIVITY OF 2 LINES SELECTED FOR HIGH AND LOW ACTIVITY IN
RESPONSE TO SOCIAL ENVIRONMENT (1 AND 3 CHICKS PER GROUP)

Line	Chicks per group		
	1	3	$\bar{X}$
High activity	1522	81	802
Low activity	81	31	56
$\bar{X}$	802	56	429

REFERENCES

Bareham, J.R., 1972. Effects of cages and semi-intensive deep litter pens
    on the behaviour, adrenal response and production in two strains of
    laying hens. Br. Vet. J. 128, 153-162.

Black, A.J. and Hughes, B.O., 1974. Patterns of comfort behaviour and
    activity in domestic fowls: A comparison between cages and pens.
    Br. Vet. J. 130, 23-33.

Boyer, J.-P., Melin, J.-M. and Bourdens, P., 1973. Activity test on young
    pheasants. Ann. Génét. Sél. Anim. 5, 417-418.

Eskeland, B., 1976. Methods of observation and measurement of different
    parameters as an assessment of bird welfare. Proc. Vth Europ. Poultry
    Conf. Malta, Vol. II, 988-998.

Faure, J.-M. and Folmer, J.-C., 1975. Etude génétique de l'activité
    précoce en open-field du jeune poussin. Ann. Génét. Sél. Anim. 7,
    123-132.

Hamburger, V., 1963. Some aspects of the embryology of behaviour. Quart.
    Rev. Biol. 38, 342-365.

von Holst, E., 1969. Zur Verhaltensphysiologie bei Tieren und Menschen
    Bd. I. Piper Verlag Munich, 1. Auflage.

Hughes, B.O. and Black, A.J., 1974. The effect of environmental factors
    on activity, selected behaviour patterns and 'fear' of fowls in
    cages and pens. Br. Poult. Sci. 15, 375-380.

Jezierski, T. and Bessei, W., 1978. Der Einfluss von Genotyp und Umwelt
    auf die lokomotorische Aktivität von Legehennen in Käfigen. Arch.
    Geflügelkunde 42, 159-166.

Jones, R.B., 1978. Activities of chicks in their home cages and in an
    open-field. Br. Poultry Sci., 19, 725-730.

McClearn, G.E., 1961. Genotype and mouse activity. J. Comp. Physiol.
    Psychol., 54, 1674.

Saleh, K. and Bessei, W., 1978. Beitrag zur genetischer laufactivität
    von wachteln. Deutscher gesellschaft für zwichtungskunde. Stuttgart,
    Hohenheim.

Savory, C.J., Wood-Gush, D.G.M. and Duncan, I.J.H., 1978. Feeding behaviour
    in a population of domestic fowls in the wild. Appl. Anim. Ethol. 4,
    13-27.

DISCUSSION

G. Martin *(FRG)*

I would like to make a point on the definition of
'activity'. You say that the animal is active when it is
feeding, when it is afraid, and so on. But the animal is
performing some kind of activity all the time; it is the
motivation of the activity which is of concern from a welfare
point of view.

W. Bessei *(FRG)*

It has been demonstrated in physiological trials that
there is an innate need for locomotor activity. If you
postulate that there is a spontaneous activity and if you put
the bird in an environment where it cannot exert this
spontaneous activity, then it may be cruel. The problem is to
measure this spontaneous activity.

G. Martin

I think it is a mistake to compare activity in cages
with activity outside. In a cage an animal is forced to
activity by the proximity of the other birds. If it is in a
free environment then its activity is spontaneous.

K. Vestergaard *(Denmark)*

I think you must admit that locomotor activity is a very
crude measure of behaviour. It has been proved that there are
many motivating influences contributing to locomotor activity.

W. Bessei

I was concerned to demonstrate a method to measure the
innate, the genetically fixed, behaviour. This can be done
with other behaviour traits. I chose locomotor activity
because it is easy to measure. You could do it with dust
bathing and other things, you can estimate the heritability,
and the genetic and environment effects. It has not been done

yet because it requires observations on a lot of individual birds.

C.C. Brantas *(The Netherlands)*

I agree with the remarks of Mrs. Martin and Dr. Vestergaard. It is very crude to speak about locomotor activity in general. You are mixing together dust bathing, scratching, searching for food, walking for fun, and so on, also pacing - and I cite Dr. Duncan - pacing is symptomatic of frustration. You are mixing together happiness parameters with frustration parameters.

W. Bessei

As regards the factors which influence locomotor activity, egg laying has been included in our statistical model so it is reasonable also to include pacing. But you cannot measure all the factors; dust bathing was not measured, we only measured the locomotion - the walking. I agree with Dr. Duncan that the stereotype pacing may be a sign of distress but I cannot accept that normal pacing must necessarily be a sign of happiness in the hen.

B.O. Hughes *(UK)*

I would like to make two points. First of all I agree with Mrs. Martin that one must try to specify what the motivation is for each kind of locomotion. But there may be a disagreement here because of the different ages of the birds involved. If one looks at small chicks it is very difficult to decide what is the motivation of their movement. Perhaps one has to recognise some kind of non-specific locomotion in this case which may be associated with exploration or some drive of that kind. In an adult bird it is much easier to recognise the approximate cause of the locomotion.

The second point relates to my own results which you mentioned. In fact, I took one group of birds from cages and put them into new cages as controls. I took a second group,

which were on the floor, and put them in similar cages. I
think this tends to support Mrs. Martin's idea that the birds
are being affected by the interaction of the bird and the
space. What I think happened there was that the birds on the
floor, although they were a coherent social group and the
group size was not changed, they were being pressed closer
together in the cage. This change in the space relationship
between the birds was probably what caused them to remain more
active than the birds moved from one cage to another.

W. Bessei

I don't think there was this kind of effect in our trials
because I tested the birds in very large cages, 1 500 cm^2 per
bird, with birds in single cages. They were not crowded and
they had a lot of space to walk around.

M. Zanforlin (Italy)

I agree that locomotion is not a specific activity but
if we compare general activity between various species, at
least in some species there is a specific need for general
movement. I think this is very important from the point of
view of animal welfare.

W. Bessei

Yes, you can see the difference in locomotor activity if
you compare between the pheasant, the quail and several types
of domestic hen. There are differences between the various
breeds of hen but there are large differences between the hen
and the quail and the pheasant. So, I come to the conclusion
that the hen may be better adapted to a restricted environment.

J. Fris Jensen (Denmark)

With regard to locomotor activity, I would like to hear
some more comments on the genetic influence.

W. Bessei

The most important point as far as welfare is concerned
is that genetically fixed behaviour cannot be changed by the
environment. We have to give the birds the right environment.
On the other hand, if the behaviour is due to the environment,
we can possibly influence it during the rearing period. It
may be better to rear birds on litter if they are to be housed
on litter later.. Equally it may be better to rear birds on a
wire floor if they are going to be housed in cages during the
egg laying period. So, we have to separate the genetic and
environmental factors in behaviour.

J. Fris Jensen

Does that mean that a character with a high heritability
is in Category 3?

W. Bessei

Yes, I think it is a sign that the behaviour is more
genetically fixed than traits with a lower heritability.

W. Sybesma *(The Netherlands)*

If there is indeed a high heritability, 0.3 seems a good
figure because milk production is at the same level and if you
have genetically determined behaviour you can get rid of it by
breeding. Then it is not a welfare factor at all.

W. Bessei

That is another point. At the very low genetic variability in the laying hen which we have found, 0.18, it may be
that at a given variability you can select for high or low
locomotor activity.

W.F. Raymond *(UK)*

Coming back to Dr. Martin's point, surely the other birds
in a cage are part of the environment; the other birds can

modify the locomotor activity. Obviously, if the birds are packed so tight that they just cannot move then you have completely restricted locomotor activity. At the other end of the spectrum there is the very large cage in which there is no limitation. In between, there is every gradation, and therefore other birds must be defined as part of the environment. I wonder how you measure this as an environmental parameter.

W. Bessei

If there is a genetically fixed locomotor activity then it has to be expressed, even in very crowded conditions. In this case the movement would disturb the other birds. If locomotor activity can be adapted to the environmental conditions then the birds would not move so much in very crowded situations.

C.M. Hann *(UK)*

It seems to me that one interesting aspect of this selection experiment was to show how easy it is to alter the genetic tendency for high or low activity. Looking at the wide variety of existing commercial stocks, it seems probable that there is a wide diversity in their behavioural needs. This reinforces the difficulty of trying to assess optimum levels for a particular factor in an environment.

R. Moss *(UK)*

I am struck by the remarkable resemblance between Dr. Bessei's diagram in relation to locomotor activity and Dr. Hughes' diagram in relation to external and internal stimuli. It seems to me that there are the same elements in both diagrams. I would like a comment from both Dr. Bessei and Dr. Hughes as to whether they agree. Dr. Bessei said that a locomotor activity need would be a given breed in a given environment. Going back to Dr. Hughes, he talked about the possibility of the 'normal' being the normal for a breed within a particular environment. So, we are looking for something which is impossible to define in optimum terms? We have to

look across the whole repertoire of behaviour, trying to
identify individual behaviour, at the same time, as Dr. Martin
said, not confusing mixed motivational or factorial behaviour.

W. Bessei

I don't think that needs any comment, I agree. I think
the only difference between my diagram and that of Dr. Hughes
is that he did not estimate the variability of the different
behaviour traits within the lines. This is done by genetic
analysis so that you can separate and quantify the genetic and
the environmental factors. He has called it internal and
external stimuli; I think part of the external stimuli is
clear, it is the environmental component of behaviour. The
internal stimuli are very difficult because they include
spontaneous activity which is a genetic component but some of
the internal stimuli are part of the environmental component.
So, I do not think you can compare the two methods exactly.

B.O. Hughes

I would like to put my diagram on the board so that I
can clarify the differences between the two approaches:

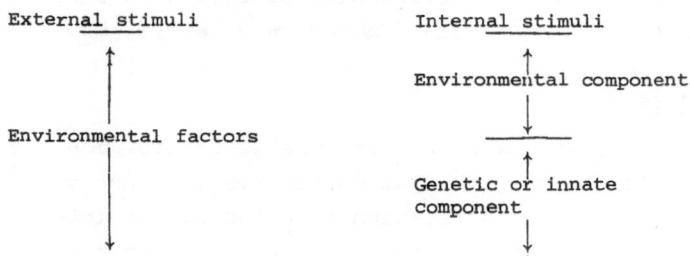

I think the important thing is that internal factors
cannot be equated with innate or inherited factors. There is
also a variable environmental component within the internal
factors.

H.C. Adler

Thank you. Unfortunately, we have run out of time. May
I thank you all for a very interesting morning.

SUMMARY AND DISCUSSION

H.C. Adler *(Denmark)*

The heading of this session is 'Behavioural and Physiological Needs". We have had four valuable papers for the
elucidation of this subject, based primarily on research and
experience in four of our countries, and indirectly, on knowledge from other countries. We have had comprehensive discussions
on the papers and now we must try to formulate some answers to
the questions posed in the title of the session - what are the
behavioural and physiological needs of egg laying hens? We
must also bear in mind the requirement to assess research and
development needs. If we cannot fully answer the questions then
that, in itself, may be an indication of where further research
is required.

I would suggest that we try to list the behavioural and
physiological needs, as we see them, under various headings,
for example: space, type of floor, material for nesting,
material for dust bathing, and so on.

R. Moss *(UK)*

I feel sure that is an approach that is acceptable to all
of us. In addition to the subjects such as space, floor type,
nesting, requirements etc. which constitute the behavioural
repertoire that Dr. Duncan spoke of yesterday, I would like to
hear some discussion on the criteria we can use to establish
that the animal within its environment is, in fact, 'in a state
of good welfare', to use Dr. Hughes' definition. I suggested
this morning that one might say the animal needs to be clinically healthy with no frustrative or distorted behaviour. Is
that the criterion for examining floor type, the need for
nesting behaviour, dust bathing, and so on?

J.P. Signoret *(France)*

There does appear to be a problem in assessing the results
concerning the various headings you have mentioned. There were

R. Moss (ed.), The Laying Hen and its Environment, 181-192
Copyright ⊙1980 ECSC, EEC, EAEC, Brussels-Luxembourg. All rights reserved.

some discrepancies in the results which may be partly due to the way the measurements were made and to interpretation. However, it is very difficult to give definitive answers to such questions as, what is the correct space for a bird? I think we have given insufficient attention to such aspects as the percentage of mortalities, percentage of cannibalism, percentage of physical injuries. These factors have an obvious bearing on welfare. Maybe we should also consider parasitism, feather loss, and so on as pointers to welfare.

From the results presented there appears to be a wide variability in the strains, in the design of equipment, and certainly in the level of management. I believe it would be of tremendous value to collect data on a large scale in the field, possibly it would be a rather crude study, but on very large numbers of animals over a wide geographical area. This would give a large number of objective results and might help to overcome the problem of strain differences, equipment differences and human management variations. After such a widespread enquiry we may be able to show for example, that a given strain housed in given conditions, gives rise to a given death rate. We can then make comparisons and hopefully pinpoint the areas where the real problems lie.

H.C. Adler

I have considered this aspect myself but since it is not included on our agenda I have assumed that the Commission already has the answers without asking us. Is that correct Mr. Moss?

R. Moss

That is correct, Chairman. The health side is important and has to be taken first. Health and welfare go together; I have preached this for a long time. Admittedly we do have areas in health where we need to investigate the disease situation but at least we are moving there. We have a lot of information and I think that can be set on one side at the present time.

The behavioural repertoire and the reaction of the animal to
its environment is the most important thing here. It may
reflect on the disease situation because the stress of the
animal's environment may predispose it to a disease. There are
problems there. However, when writing the agenda, 'behavioural
reaction to environment' was the thing to be concentrated upon.

H.C. Adler

Thank you. So, although we would all agree that these
parameters are certainly important in the evaluation of a
welfare situation, they are not on our agenda here.

B.O. Hughes *(UK)*

Mr. Chairman, I think the items you have listed are all
important but I think we might add one or two more. In relation
to this morning's contribution from Dr. Bessei, as well as
locomotion we have to look at the converse of this. It was
striking from my own work, that in cages, hens spend a lot of
time standing still and doing nothing. This was not the case in
other types of environment, and yet they spend very little
time, laying down and resting, in cages. Perhaps we should add
'resting' to the list as a possible behavioural need that
should be considered.

H.C. Adler

Yes, thank you. If there are no further suggestions may
we now consider the various headings, starting with space. As
a starting point I would like to quote from a working paper
presented by the British delegation to the Standing Committee
of the Council of Europe at a meeting in October.

"Recommendations for specific stocking densities are
difficult to set out because of the many factors which
have to be taken into account. In addition to consideration of the usable space for each bird, consideration
needs to be given to the heat output of the birds in
relation to the stocking density within the house as a

whole. If the ambient temperature is likely to rise
above 25°C, the stocking density will need to be reduced
unless the temperature can be controlled by a ventilation
cooling system. Conversely, minimum densities are more
important when attempting to maintain optimum house
temperature in cool or very cold climatic conditions.
Stocking density should also take into account the size,
weight, behavioural and functional characteristics of the
birds. However, it would seem to be precipitate to
specify definite parameters, except perhaps for extremes,
for any of the environmental measurements."

May I have your comments, is that what we feel?

B.O. Hughes

I am not sure what the role of the meeting is at this
stage. I feel that yesterday there was a certain amount of
acrimony when we tried to deal with the problem of space. It
is clear to me that there was no agreement between us. All I
can do now is to reiterate what I said yesterday but it does
not become any more true if I say it twice.

R. Moss

If, there is disagreement, Chairman, within the body of
colleagues here, then it should be known, it should be
illustrated, it should be emphasised. This is an area in
which there ought to be agreement on how that disagreement is
resolved.

H.C. Adler

There certainly is disagreement on the question of space
requirement, we have had everything mentioned, from one hectare
down to almost nothing.

B.O. Hughes

Mr. Chairman, in order to resolve the impasse, one could make a population survey and so draw up a histogram. If people were only allowed to name one figure we could see what the population spread is, whether there is any kind of mean or whether it is random.

R. Moss

I am afraid that Dr. Hughes and the others are going to have to be put on the spot here. May I come back, yet again, to my suggested criteria for well-being: 'total clinical health with no frustrative or distorted behaviour' and qualify it in accordance with Dr. Bessei and others with 'for a given breed in a given environment'. Are there colleagues here who would be prepared to give an opinion on that basis? Dr. Duncan has talked about social space; what does be believe social space should be in minimum terms, in accordance with these criteria?

C.C. Brantas *(The Netherlands)*

When we agree with Mr. Moss it will be the end of poultry keeping! I do not think it is the task of we scientists to give criteria; we can only give facts. We can only say when the animal has one hectare of space there is very little frustration; when it has 600 cm^2 the level of frustration is so high, in 400 cm^2 it is so high. It is the task of the politicians to choose; we can only give the information.

R. Moss

But Chairman, we already have there a statement which is, in fact, part of the criteria which others may not agree with. That is that with one hectare per hen there is virtually no frustrative behaviour.

I.J.H. Duncan *(UK)*

In one hectare of space there may be many other factors affecting the welfare of a bird. If that one hectare happens to be outside and there are birds of prey about, the hen could be very frightened. Frustration is only one aspect of welfare; we have to consider all the other aspects as well.

I have been put on the spot and asked to suggest a figure. Animals do have a personal field and if these personal fields come into contact, social friction results. Animals will try to avoid their personal fields overlapping. I am disappointed that no one has measured this but from the information we have at present I think it might be something like 600 or 700 cm^2 per bird if the group size is fairly small, but these two things will probably interact.

J.A. Hill *(UK)*

I presume you are talking specifically about a cage system. A litter system, for example, would be a different ball game altogether because of the increased potential for social interaction in a larger group size.

I.J.H. Duncan

Well, no, I am taking into account the group size. The theory has it that these personal fields will vary in size according to the activity that is being performed by the animal, but it does not take account of differences in environment, wire as opposed to litter.

B.O. Hughes

I agree it is a political decision. I think perhaps I should reiterate what I said yesterday. It seems fairly clear that it is a continuum, that birds will benefit from increased space but the benefit will decrease progressively. It is obviously an exponential design reaching asymptote so that as you add further increments of space, each time the benefit will become less. However, the quality of space is of concern as well as the quantity. It may be better to give the bird a small area of wire and use the money you have available to provide some other facility, such as an area of litter for dust bathing. This is a political decision. All we can say

is that increased quantity and quality of space will benefit
the birds and we cannot say which will benefit them most until
more work has been done in this field.

C.M. Hann (UK)

In terms of pay-off between space and other factors in
the environment, a progressive reduction of birds in a given
area reduces the output of heat. One could reach a stage
fairly quickly in the northern part of Europe where by reducing
numbers not so very greatly, the winter time inside temperature
of the house would fall possibly to near freezing point. This
might be too low; it may be considered that this is too great
a sacrifice in return for the extra space. So there may be a
pay-off between these two factors, for example.

G. Martin (FRG)

It is not possible to weigh clinical health parameters
against ethological parameters, both are necessary for the
well-being of the animal.

J.P. Signoret

I think we all agree with that. It is just that there is
no point in considering ethological parameters unless a bird is
clinically healthy. If a bird is diseased or injured then it
is quite obviously not in a state of well-being. This was the
reason for my suggestion for a wide survey, perhaps on 10 000
hens spread over 100 different farms. This would give an
indication if there are particular types of husbandry and
management methods which give rise to disease and injury. If
these could be eliminated then at least there would not be any
welfare problems arising from clinicial parameters.

C.C. Brantas

I think it is a mistake to speak of welfare and non-
welfare. No animal in the whole world has total welfare, that
is Utopia. It is impossible to reach a situation where an

animal is not frustrated, even in one hectare! We have to look
for a compromise - a little bit of physical injury, a little
bit of behaviour frustration, and so on - that must be acceptable
for the politicians.

L. Spanoghe (Belgium)

In concentrating on welfare it is possible to reach
conclusions which are against health. For example, if you
put a sand bath in a battery system, for the first hour you
have sand and then you have litter. Litter can be the source
of parasitic and contagious diseases. So there can be conflict
between welfare and health.

H.C. Adler

Have you got the information you expected from us on the
problem of space!

R. Moss

I have received the information I expected but not
necessarily the information for which I had hoped. I would
like to ask a final question of Dr. Duncan. He has talked of
social space and knowing him very well I believe he must have
given some thought as to how it can possibly be defined in
various systems, or at least in one system. There is work
which I am sure Dr. Duncan is aware of that gives indications
of the way towards defining the social space of the laying hen.

I.J.H. Duncan

Yes, I have one or two vague ideas about how these
measurements could be made. We must remember that although
there are forces which are spreading hens out, the domestic
fowl is a social animal and there are also forces pulling hens
together. If you put half a dozen hens into a room this size
they would not space themselves out to gain a maximum area,
they would form a small flock and move about together. One
could conduct an experiment like that, by having a virtually

unlimited space, looking at small numbers with overhead cameras, increasing the numbers, looking at social interactions during different activities, and so on. In that way one might eventually be able to measure the forces that draw the hens together as opposed to those which push them apart.

G. Martin

Mr. Moss, if I understood correctly, you are asking again for a figure to be put on the space requirement of the hen.

R. Moss

There have been indications from Dr. Bessei, Dr. Hughes and others that for locomotor activity, for example, and other activities, there is surely, within a given breed and in a given environment, a particular parameter.

G. Martin

I want to stress what I said earlier, that you cannot consider quantity of space without also considering quality of space. Also, space requirement varies for the different behaviour patterns.

I.J.H. Duncan

Yes, I agree that you have got to take account of the behaviour patterns that are being shown. For dust bathing it appears that hens do not space themselves at all; they can come into close physical contact. If we were only to consider dust bathing we could say that hens only need the physical space taken up by their own bodies, but it is different with feeding behaviour where they need a little more space - and so on. We would need to work through the repertoire.

P.M. Schenk *(The Netherlands)*

Perhaps the two extremes - sleeping, on the one hand, when the birds are fairly close together, and nesting behaviour, on the other hand, when they try to isolate themselves from the flock for a time but return later.

I.J.H. Duncan

Nesting is an interesting case because it may be possible to overcome that artificially by cutting the other birds off visually. It may be that the bird does not actually need more space but requires visual isolation. We don't know.

B.O. Hughes

Mr. Moss asked about possible research. In this connection the further work that I mentioned in my talk is of interest in that hens that have been allowed to choose do give a lot of information. As Dr. Duncan has said, when birds are in a flock with adequate space they are choosing in the sense that they are spacing themselves out. However, in my studies, in one case they tended to choose a large space rather than a small space - about 1 - 1.5 m^2 as compared to about 800 cm^2. But that only looked at what the hen was doing when on its own. One could incorporate this kind of approach to looking at the requirement for space under other circumstances. For example, one could have hens entering different areas to carry out specific behaviours. There would be a central area which they could leave for other activities: dust bathing, nesting, feeding, contact with other birds, or any other locomotion or need. They would be able to choose these in turn, return to the central area and then go out again. By manipulating the various amounts of space available, one could see how far they had to be decreased before one affected the usage of that space.

W. Sybesma (The Netherlands)

Listening to the discussion, I believe the argument could go on ad infinitum and still not reach a conclusion. I would suggest that it might be best to take one specific environmental situation at a given time. Then in our discussions in the next session everything can be related to that, either as better or worse. In this way the scientists can provide information for the politicians.

H.C. Adler

Mr. Moss, does that mean that we can go on to floor type now and leave space for tomorrow?

R. Moss

I think that is so, Chairman. One of the things that is apparent from this discussion is that there is a certain amount of blindness amongst us all as to exactly where we might wish to go; that each of the scientist colleagues are pursuing a particular line. The totality of it all, as Dr. Sybesma has said, will come together in the individual system when you look at all the parts. Perhaps then we can separate it out into what is necessary to look at. Certainly there have been indications today and I would hope that Dr. Duncan will continue to think about his potential experiments and research into social space.

H.C. Adler

Really it is time to finish Session II. Are there any more remarks before we do so?

P.M. Schenk

I would just like to go back to the question that was raised in the discussion this morning, the question of calls. I think it is very important to know why the calls get certain names. Dr. Duncan objected to Dr. Fölsch referring to a call as the 'friendly' call. I think there are three distinctions to be made. Firstly, we can refer to a call simply as twittering, peeping and so on. Secondly, we can try to express the causation of the call in the name, but first we must know something about the causation. Thirdly, we can interpret the function of the call as a social signal and indicate that in the name of the call, for example 'appeasement' call. When Dr. Fölsch referred to the 'friendly' call he was saying something about the function of the call. He had indications that the function of that call is to communicate

with other birds, that it does not have an attacking motivation,
and so I think he is right to call it a 'friendly' call or a
non-aggressive call. I think we should think about this before
we object to the names of calls. If we do distinguish calls
in this way, we can observe them in various conditions and
environments and draw valid conclusions as to whether a
particular environment is stimulating agression, or whatever.

I.J.H. Duncan

I fully agree. Dr. Fölsch was using the name of a call
given by Baeumer. I am not satisfied that Bauemer did sufficient
work into the causation or function of this particular call to
name it 'friendly'.

H.C. Adler

Thank you. I must now close Session II.

SESSION III

MEASUREMENT OF ESSENTIAL AND BEHAVIOURAL NEEDS
AS PROVIDED BY THE PRESENT HUSBANDRY SYSTEMS

Chairman: W.F. Raymond

MEASUREMENT OF ESSENTIAL AND BEHAVIOURAL NEEDS AS PROVIDED BY PRESENT HUSBANDRY SYSTEMS: BATTERY, 'GET-AWAY' CAGE, AVIARY

R.-M. Wegner

At our Institute for Poultry and Small Animal Research in Celle we have, for 2 or 3 years, been working on the possibility of keeping laying hens for egg production in intensive systems which provide the birds with an environment to satisfy their behavioural needs as well as satisfying the economic needs of the producer and consumer. Three years ago we started investigations in three main areas:

1. Comparisons of the systems – deep litter with range,
   deep litter without range,
   cage batteries

   with regard to: behaviour
   performance
   mortality
   egg and meat quality
   anatomical, clinic-chemical
   or hormone-physiological traits
   working together with several other research stations.

2. Investigation of the 'get-away' cage system.

3. Investigation of the aviary system.

With regard to comparison of systems we started in 1977 with 2 300 light hybrid laying hens divided into 12 groups, 4 replicates per system. We are now running the third experiment, having finished two experiments of 52 weeks each, starting with hens at 20 weeks of age. In the first two experiments we found significant differences between the systems with regard to the following traits:

R. Moss (ed.), The Laying Hen and its Environment, 195-205

a) egg weight - higher in cages

b) mortality - lower in cages

c) no losses by cannibalism in cages as found in deep litter without range

d) no losses by wild birds as found in the deep litter system with range

e) feed consumption lower in cages, feed conversion - better in cages

f) percentage dirty eggs - lower in cages

g) yolk colour - darker in cages

h) body weight at slaughter - lower in cages.

No differences were found in the blood-chemical analyses and the hormone-physiological tests.

Concerning the investigations into behaviour - the main subject of this experiment - observations are made during 3 weeks per quarter, 3 days per week, 4 hours per day, 1 hen per hour, simultaneously in the 3 systems. Forty eight different behaviour traits are being observed, of social behaviour, of comfort and resting behaviour, of feeding behaviour as well as standing or walking and sitting or lying. The duration as well as the frequency of those traits are registered. Altogether 144 different hens per system and year are observed.

In addition, special observations take place with regard to nesting and egg laying behaviour, also during 3 weeks per quarter, 35 different traits are observed of 55 different hens per system and year.

As you may see we are registering a very large amount of data which means that statistical analyses are very difficult and complicated and therefore not yet finished from the first 2 experiments. Two statistic experts are working on these data. Differences have been found, of course, in several of the behaviour traits, similar to those in other research stations.

During the 3rd experiment, which will run until September 1980, we included a medium-heavy hybrid, laying brown shelled eggs, as well as the same light hybrid which we tested during the first two experiments.

Up to now the medium-heavy hybrid has shown quite a different behaviour from the light hybrid. It seems it is necessary to test other light and medium-heavy hybrids before being able to present repeatable results with regard to significant differences in behaviour traits between the systems. The evaluation of all these results will be an especially difficult task.

The experiments with the 'get-away' cage, developed in the United Kingdom, started in 1976, because some ethologists in our country thought it could be an alternative to the conventional battery cage, and provide for more of the behavioural needs of the laying hens. We have modified the English type of 'get-away' cage using three different variations.

Variation 1 was 80 cm high, 100 cm wide, 65 cm deep; and had 13 perches, 2 feed troughs; it was tested with 16 or 20 hens per cage; and had 4 nests, litter nests or roll-away nests with plastic mats at the nest floor.

I started the first experiment with this variation using about 3 300 laying hens of two different light hybrids, 16 hens per cage, in autumn 1976 in Krefeld-Grosshüttenhof whilst I was at Bonn University, before I took over the Institute in Celle in October 1976. It lasted about 17 months and gave the following results:

'Get-away' cages with litter nests showed a significantly lower number of eggs produced with a high percentage of dirty eggs and destroyed eggs, because hens used the nests not only for laying but for scratching and dust bathing. There were high labour costs for collecting the eggs and filling the nests with litter. In the 'get-away' cages

with roll-away nest we found only a slighly higher percentage of dirty eggs and no difference in egg production in
comparison to the conventional cage type. The behaviour
of the hens observed by Brantas and co-workers was better
in the 'get-away' cages.

The second experiment, again, with more than 3 000 hens
of one light hybrid, and lasting again for 17 laying months was
completed in January 1980. We compared 16 and 20 hens per
'get-away' cage. We used only the roll-away nest and tested
different kinds of mats in the nest as well as a dust bath
beside the nests. We lost so many eggs in the dust bath that
we took it away after a few weeks.

The results of the second experiment are not yet ready
for presentation. Again the behaviour was observed by Brantas
and co-workers.

Variations 2 and 3 of the 'get-away' cage type were
developed and tested in our Institute in Celle.

> Variation 2 was 80 cm high, 100 cm wide, 100 cm deep; and
> had 4 perches, 2 feed troughs; it was tested
> with 20 - 25 or 30 hens per cage; and had
> 4 nests, roll-away nests, and 1 dust bath.

> Variation 3 had the same construction but was only 55 cm
> high, had 3 perches and was tested with 15,
> 20 or 25 birds per cage.

The first experiment with 4 000 light hybrid hens started
in August 1978 and lasted one year to August 1979. The second
experiment started in September 1979 and will be finished in
September 1980.

The results of the first experiment with both variations,
with 4 replicate blocs per cage variations and per number of
birds per cage, compared with 3 and 4 birds in the conventional

type of cage, were the following:

Slightly lower egg performance in all 'get-away' cages, probably caused by egg losses from destroyed and eaten eggs in the dust bath; slightly higher mortality in the 'get-away' cage types, probably caused by difficulties for the birds at the beginning of the experiment in finding the nipple drinkers on top of the cages. The differences were not significant. There were no differences between the number of birds per cage in performance traits or mortality. That means with two tiers of the higher density 'get-away' cage type with 30 birds per cage, or with 3 tiers of the lower density 'get-away' cage type with 20 birds per cage it would be possible to keep the same number of birds per m^2 of poultry house as with the conventional type of cage.

However, this type of 'get-away' cage has still 2 main disadvantages:

1. It is very difficult to look into the cage, to control the birds, to discover sick or dead birds.

2. In contrast to our first tests of the behaviour of the hens in this type of 'get-away' cage, published by Brantas and others in 1978, the dust bath is not only used for dust bathing, but for egg laying also - a great disadvantage from the economic and hygienic viewpoints!

We are trying now to eliminate these disadvantages by modifying the floor of the nest to make it more attractive for the birds for egg laying than the dust bath. We are also testing different hybrids - light and medium-heavy - in these two 'get away' cage variations.

We are not sure at the moment whether it will be possible to overcome the following disadvantages of the dust bath:

a) eggs laid in the dust bath

b) too much sand removed from the dust bath into the cages and dropping pit

c) high labour costs to fill and clean the dust bath.

Up to now we have not had sufficient time to make intensive observations of the behaviour of the hens in the Celle 'get-away' cage types because our three scientists and three technicians, working in the behaviour field, have had time only for our first project, comparing the different systems, which requires a very high amount of time and labour. This project must be completed at the end of this year. Then the financial support for two scientists and two technicians will end. I hope to receive further financial help from one or other source, otherwise we will have to finish all our behaviour observations altogether.

Experiments with regard to the aviary system started in 1978. This system, also developed in the United Kingdom, is characterised by using the free room between the floor and the ceiling of a poultry house by installing other levels of platforms or perches at different heights above the floor. With this system it should then be possible to increase the number of hens per m^2 floor area, to raise the temperature during winter time, to save feed and energy, to lower the house and equipment costs per hen, and thus to decrease egg production costs in a deep litter house, without having an adverse effect on the behaviour of the hens, perhaps reducing the danger of cannibalism by giving the hens a greater chance to escape.

We started our experiments with the aviary system in, for poultry keeping a new type of house, a low-cost house, a so-called 'Folienstall' that means a half-round plastic house used in nurseries and tested by our colleagues at the Federal Research Station for Agriculture in Braunschweig for rearing cattle. The first version of this house that we used was not insulated and had no forced ventilation. The house was 8.5 m wide and 15 m long; 3 m of the length were used for installing

the chain feeder machines and the egg collection facilities of
the 'Farmer Automatic Nest'. The floor area inside the house
was about 100 m^2. We started with 1 000 light hybrid hens, that
means 10 hens per m^2. Instead of platforms at different heights
as used by Elson in his aviary system in the UK, we used equipment similar to the perches used for cockerels during rearing.
Of course it is necessary to provide enough feeder space for
the hens at the different levels. We installed one feed trough
with chain on the ground and a second one on the highest perch
area. The water was provided by nipple drinkers. As nests we
used the 'Farmer Automatic Nest' with buckwheat hulls as litter.
The eggs are transported together with the litter out of the
house to a separate room where a fork separates the eggs from
the litter for easy collection.

We have finished the first experiment with this aviary
system after a 15 month laying period. Performance data were
similar to those with the conventional density of 6 hens per m^2.
Disadvantages were: we had not installed a dropping pit below
the perches which caused the hens' feet to become dirty and a
higher percentage of dirty eggs. The nests were used very well
by the hens; the percentage of eggs laid into the litter was
lower than 1%.

One month ago we started the second experiment with this
aviary system. We built a second plastic house in the meantime
and we made some changes in the construction and the equipment.
We drew a second foil above the first one with insulation on
top of the hall to avoid water dropping down in the winter
months. Furthermore we installed two dropping pits below the
perches to reduce the percentage of dirty eggs. We also
equipped the house with two ventilation fans drawing the warm
air of the house, produced by the hens, between the two plastic
walls to increase the insulation effect.

We are now testing two different numbers of hens per m^2,
10 or 15 hens per m^2. We also have a kind of control group
with 6 hens per m^2 in a conventional type poultry house.

Unfortunately, up to now we have not been able to observe the behaviour of the hens, we can only say that we had no special problems with cannibalism during the first experiment.

In conclusion we have to say that we still need to continue our experiments with this aviary system, as well as with the 'get-away' system, before we are able to recommend it to the poultry farmer.

Besides these 3 main subjects we have started some smaller experiments with regard to welfare, these are:

a) registration of sounds of hens in different systems by telemetry

b) behaviour of hens during darkness in different systems using infra-red photography

c) influence of the rearing system, cage or litter rearing, on the behaviour of laying hens in cages

d) dust bathing and comfort behaviour of hens in different systems

e) behaviour, performance and mortality of hens kept continuously for at least 4 years, in cages or on litter

f) duration and intensity of restlessness before laying of hens in cages using video-techniques.

As you can see we have started to investigate quite intensively the behaviour and performance traits of laying hens in different systems but the longer we work in this field the more we realise that we are only at the beginning and that it will take many more years to be able to present repeatable results which may help to decide whether the different keeping systems are acceptable to the laying hen, as well as to the people working in them and with them.

DISCUSSION

R. Tauson *(Sweden)*

I would like to ask Dr. Wegner four questions. Have you made any measurements of ammonia in the different systems, or are you planning to do that? I would also be interested in the percentages of relative humidity in the different systems? Thirdly, how often did you change the litter in the systems? Lastly, what was the feed consumption during these rather cold days and nights?

R-M. Wegner *(FRG)*

We did measure the ammonia content in the air of the poultry house in all the experiments. We found very high values in the winter time in the deep litter system in comparison with the other systems but it was not higher than 50 ppm which is the maximum permissible level in our country.

On you second question, the relative humidity in the deep litter system in winter was as high as 90 - 100%.

We did not change the litter during the laying period in any of the systems.

With regard to feed consumption, with the experiment comparing the different keeping systems we had a feed consumption per bird per day of 135 g in the deep litter system with the range. We had a feed consumption of 134 g per bird per day in the deep litter system without range. In the battery system it was 122 g per bird per day. In the aviary system, during the first experiment, feed consumption was about 126 g per bird per day.

W.F. Raymond *(UK)*

Dr. Wegner, I hope you will be publishing some figures in your paper, some data on egg production and so on.

R-M. Wegner

We have not published any figures so far because we thought it would be better to wait until we had finished some more experiments, but we will think about it.

W.F. Raymond

Would it be possible to publish some figures even if you indicate that they are preliminary ones? It would be very valuable to this group to have some quantitative indications in addition to the qualitative indications you have given us.

R-M. Wegner

We will think about this; if possible I will do it.

L.H. Huisman *(The Netherlands)*

You were talking about getting eggs in the dust bath in the 'get away' cage. How did you solve that problem?

R-M. Wegner

We closed the dust bath! We are trying now, in the second experiment, to solve this problem. At the moment we do not have a solution; we are not sure that we can find one but we are still trying.

R. Tauson

You mentioned that you had obtained some effect by putting light over the dust box.

R-M. Wegner

Yes, that was one possibility we tried; it helped a little but not enough to make us recommend this method. We are trying something else now. Besides the large poultry house with the different types of get away cages we have four single get away cages where we can move all the pieces of equipment around the cage. We are trying putting the nest on the lower tier and the dust bath above the nest. We are moving the perches far enough away from the dust bath to make it difficult for the birds to reach it; in this way we hope to overcome the problem of eggs being laid in the dust bath. However, this is just a first experiment, we still have to prove that this would be a way to avoid the birds laying eggs in the dust bath.

W.F. Raymond

If there are no more immediate questions I suggest we go on now to the next paper because I suspect there will be quite a lot of interaction between the papers in this section.

PUTTING SCIENCE INTO PRACTICE

J. Amanda Hill

FLOOR HOUSING SYSTEMS FOR COMMERCIAL EGG PRODUCERS

Any consideration of research and development needs must
be based on an assessment of past work and the level of current
knowledge. When the discussion appertains to welfare, it must
also take into account possible differences between laboratory
conditions and commercial practice. Because of pressure from
welfare lobbies, governments will eventually be forced to make
decisions, and possibly to introduce legislation. It is not
only important that these decisions are based on scientific
evidence, but also that the evidence is appropriate to the
context.

On occasions there may be conflict between that which
should be provided on scientific grounds and that which can be
provided in practice. The latter should not be ignored. Any
husbandry system, whether it is chosen for economic reasons,
welfare reasons or because it is a compromise between the two
extremes, must be manageable. The welfare of the bird is more
heavily dependent on good management than on any other single
factor. I make no apologies therefore for introducing an intangible into a scientific argument.

The purpose of this paper is to consider some aspects of
behaviour which have been investigated experimentally, to
discuss whether the results are applicable to commercial floor
housing systems, and to outline those areas where more work is
required. In the latter case both scientific and husbandry
problems will be considered. In the paper a floor housing system
is taken to be any non-cage system in which the birds are permanently indoors within a controlled environment. This will

R. Moss (ed.), The Laying Hen and its Environment, 207-225
Copyright © 1980 ECSC, EEC, EAEC, Brussels-Luxembourg. All rights reserved.

include litter, any raised floor system, e.g. slats, wire or plastic floors and aviaries.

1. SOCIAL BEHAVIOUR

Hens live first and foremost in a social world, and as explained by McBride (1970), the peck order is the means by which social behaviour is controlled. The husbandry system affects social behaviour by virtue of the constraints that it puts on movement. Obviously, a floor housing system permits greater social interaction than a cage system.

In most welfare debates certain aspects of social behaviour, such as aggression and the effects of stocking density, receive the most attention, and it is these that will be considered today.

(a) Stocking density

The important question of whether crowding is stressful to a bird does not appear to be easily answered. Experimental evidence is conflicting, although there is a tendency for reports to describe greater behavioural disruption and more signs of physiological stress with higher densities and large group sizes than with lower densities and small group sizes.

The literature describing the effects of stocking density on the production parameters of caged layers has been reviewed by Hughes (1975). As pointed out by this author, in much of the early work colony size and area per bird were confounded. When the two variables are separated most of the evidence suggests that increased colony size depresses egg production, raises food consumption and increases mortality. Decreased area per bird depresses egg production, reduces food consumption, lowers body weight gain and increases mortality. The effects are independent and additive. Certain strains appear to tolerate crowding better than others, but such differences are not consistent upon repetition. Such information is readily

translated into commercial practice and many UK egg producers
have moved towards small colony sizes of 4 or 5 birds in a
508 mm wide cage.

Comparable information from flocks housed in floor systems
is scant. Furthermore, the data that are available are difficult to interpret in commercial terms because of group size
differences. Experimentally, group sizes ranging from 10 to
100 birds are frequently used, and although space allocations
may be commercially representative such group sizes are
certainly not. In a commercial floor housing system a 'pen'
may contain anything from 500 to 5 000 birds. In these circumstances what is the 'effective group size'? How far do individuals move and how many of the birds do they come in contact
with? McBride and Foenander (1962) and Craig and Guhl (1969)
suggested that individuals do not range over the entire area
that is available to them, but restrict their movements to a
certain portion of the space occupied by the flock. Obviously,
such behaviour has implications for the number and allocation of
feeders, drinkers, nest boxes, etc. However, Hughes et al.,
(1974) found that the movement of 32 tagged individuals within
a group of 600 birds lay on a continuum ranging from apparent
randomness at one extreme to non-randomness at the other. In
all their experiments most of the marked birds were sighted in
all areas of the 13.42 x 5.20 m house, indicating that their
movements were incompatible with McBride's idea of a home range.
Hughes et al. ascribed the difference in results between their
experiments and those of the other workers to strain differences
in aggressiveness and the effects of light intensity. A similar
exercise at Gleadthorpe Experimental Husbandry Farm supported
the view of Hughes et al. that the movement of individuals is
very variable. One hundred birds were tagged with numbered wing
tags in a pen containing 450 Warren ISA egg layers. The pen
measured 9.75 x 3.0 m, but had two tiers of slats, giving a
total floor area of 43.55 m^2. The results could not be said to
be definitive, observations being limited to 8 x 2 hour observation periods. However, they did suggest that some individuals
covered the whole area of the pen whereas others were observed

in only one or two areas. Approximately 40% of the marked birds
appeared to remain on one tier level.

(b) Aggression

Contrary to popular belief, several authors have demonstrated that there is less aggression in cages than in litter
systems, e.g. Craig and Bhagwat (1974), Polley et al., (1974),
Bareham (1972) and Hughes and Black (1978). By observing the
changes in agonistic behaviour seen when birds were transferred
from pen to cage and from cage to pen, the latter authors concluded that this is because in cages subordinates are more
firmly under the inhibiting influence of the dominant bird. In
view of this, any demonstrable relationship between levels of
aggression and egg production would assume importance because
it would predict that production in cages would be superior to
production on litter.

Hughes (1977) and McBride (1964) could find no relationship between agonistic behaviour and egg production in caged
layers, although James and Foenander (1961) found that social
rank was related to age at sexual maturity, but not to the
subsequent rate of lay. In the latter two experiments birds
were housed in individual cages and aggression was scored in
paired encounters. However, Hughes (1977) used a technique
more appropriate to the commercial context. He administered a
fat soluble dye to each bird so that their eggs could be identified whilst they remained housed in colony cages.

In contrast, there have been several reports of a link
between social status and egg production in flocks kept in floor
pens; for example, McBride (1964), Biswas and Craig (1971),
Guhl (1953) and Tindell and Craig (1959). Tindell and Craig
and Biswas and Craig each found that age at first egg was
negatively correlated with social status. Thus, if part production records are used, there is likely to be a high correlation between rate of lay and egg production which may not be
evident if data is collected for a complete laying year. Biswas

and Craig (1971) showed that social rank and social tension index (number of aggressive acts/number of submissive acts) were significantly correlated with the number of eggs laid from 18 to 37 weeks, but that there was no correlation between these variables from 38 to 57 weeks. However, in this case the effect on sexual maturity was sufficiently large that when the entire period from 18 to 57 weeks was considered the relationship was still apparent.

One interpretation of this rather limited evidence is that due to increased social interaction and possibly an associated increase in competition, sexual maturity and consequently, for a given recording period, rate of lay may be adversely affected by floor management systems. Experimental evidence, using comparatively small group sizes, does not always support this view. For example, Gowe (1955), Lowry et al., (1956), Johnson (1964) and Christmas (1974) reported that hens housed on the floor lay more eggs than those in cages, but other workers, for example, Miller (1956), Bailey et al., (1959) and Shupe and Quisenberry (1961) reported that the reverse is true. Differences in the flock sizes and the strain of bird used may partially account for the discrepancies. Commercial evidence, on the other hand, tends to support the argument. A British Egg Marketing Board survey for 1968/69 shows the average hen day egg production to be 202 eggs per bird on floor systems compared to the figure of 230 for cages. More recent figures are hard to come by as about 95% of all commercial egg layers are now housed in cages. It cannot be denied that the performance of flocks in both cages and floor housing systems is often vastly superior to this. The great variability in commercial performance must be partially attributed to that intangible factor, management.

McBride (1960) postulated that the relationship between dominance and productivity was curvilinear and that under good management fewer birds in the flock are affected. The aspects of management involved would be those that allow competition to occur. Guhl (1953) suggested that birds high in the peck

order have precedence at the food trough, nests, etc. However
this may be an oversimplistic view. A dominant bird may not
be dominant in all situations, and as pointed out by Hughes
(1977), a dominance hierarchy obtained by recording aggressive
interactions may not correlate well with a competitive order
where hungry or thirsty animals have limited access to food or
water.

2. FEEDING BEHAVIOUR

Savory (1980) has recently reviewed the literature
related to diurnal variations in feeding behaviour. Although
five patterns were described, most authors agree that the
laying hen exhibits a pattern characterised by a peak during
the latter part of the light period. Various causal factors
have been investigated but the reproductive state of the bird
appears to be one of the most important single factors determining this pattern. This raises the question of whether it
is an important requirement that all birds should be able to
feed at once. In conventional battery cages (510 mm wide x
460 mm deep) housing 5 birds, all the birds cannot feed simultaneously. Thus it is possible that some birds are prevented
from feeding at peak feeding times. To support this view,
Hughes and Black (1976) have shown that in shallow cages (610 mm
wide and 305 mm deep) where food-trough allocation is increased,
the peak in feeding activity during the latter part of the day
was accentuated. Duncan and Wood-Gush (1972) showed that
experimentally thwarting feeding behaviour resulted in displacement preening and stereotyped pacing movements, the latter
being symptoms of frustration. If some birds become frustrated
because they cannot feed when they want to, should it be
interpreted as a welfare disadvantage or normal social
behaviour?

If it is considered that all birds should be able to feed
simultaneously, the means by which they can do so must be provided. In a floor housing system this is difficult. Firstly,
as pointed out earlier, there is variability in the distance

that birds travel and it is doubtful if all feeders are used equally. Secondly, the type of feeder that is used may influence the amount of feeding space required. Commercially it is believed that by providing a chain feeder you are doubling the space available to the birds because they can feed from either side. However, as explained by McBride (1970), dominance or submission is only expressed when a bird is within a certain distance of another known as the bird's personal field. These personal fields do not have an equal radius in all directions, but are greater directly in front of the face. Thus it is highly unlikely that birds would stand directly opposite each other at a chain feeder. Tube feeders, on the other hand, have a central portion which prevents birds on opposite sides of a feeder from seeing each other.

Deciding on a suitable trough space allocation is complicated still further in floor systems which allow birds to perch on the feeders. The effective feeding space can be reduced by half on a chain feeder which has not been fitted with anti-perch wires.

Competitive effects may be heightened in floor housing systems because the requirement for food may be greater. Due to lower house stocking densities in floor systems it is not possible to maintain house temperatures comparable to those in commercial cage units. There is ample evidence to show that food intake rises $1\frac{1}{2}\%$ for every $1^{\circ}C$ fall below the temperature of $21^{\circ}C$ which is recommended in the UK. Furthermore, the increased activity in floor systems will also result in an increased energy requirement.

3. DRINKING BEHAVIOUR

Similar principles can be applied to drinking as have been applied to feeding in the previous section. Laying birds show diurnal variation in drinking behaviour, the pattern being very similar to the pattern of feeding. Practical considerations in the selection of drinkers may be even more

critical than in the selection of feeders. If a drinker is
chosen which allows birds to waste water, the litter, or the
droppings in the pit beneath a raised floor, will become wet
and high levels of ammonia will ensue. Twenty five parts per
million of ammonia is the maximum level which UK health and
safety recommendations allow for staff experiencing prolonged
exposure, i.e. a full working day. Fifty ppm can cause blindness in birds. The only solution is to increase the ventilation
rate which lowers house temperatures and further increases the
requirement for food.

4. NESTING BEHAVIOUR

The nesting and pre-laying behaviours of the laying hen
have been studied extensively, principally by Wood-Gush at the
Poultry Research Centre in Edinburgh. Unfortunately, those
aspects which have been studied do not lend themselves to
practical application. Wood-Gush and Gilbert (1969) and Wood-
Gush (1972; 1975) have given excellent descriptive accounts of
nesting behaviour and nest construction in litter pens devoid
of artificial nests. The physiological basis for nesting
behaviour has also been studied by Wood-Gush (1975) and the
following relationships have been demonstrated:

Behaviour	Control mechanism
Nesting call)	
)	Oestrogen
Orientation away from the flock)	
Nest examination)	
Nest entry)	Postovulatory follicle-
)	acting hormonally
Sitting)	
Maintenance of the hen's attention to the nest	Ventral hyperstriatum

The one aspect which is of vital importance commercially, that
of nest site selection, has been largely ignored. Furthermore,
the few papers which have been published on the subject have
been inconclusive and often the experiments have been ill-
designed. Variables which have been investigated include the

height of the nest, darkness and the nesting material (Wood-Gush
and Murphy, 1970) age at housing and the effect of light
(Dorminey, 1974) and the colour of the nest (Hurnik et al.,
1971; 1973a, b). One common factor to emerge from several of
the studies was the enormous strain difference in the tendency
to lay eggs on the floor. Some strains appear to be very loath
to use conventional nesting systems. The variability between
individuals has also been highlighted by Hurnik et al. (1973b).
Because of this variability between strains and between individuals, it may be impossible to provide a commercial nesting
system which is attractive to all birds. However, if hens are
to be kept in floor systems, an attempt must be made to establish the basis for nest site selection and subsequently to
develop a nesting system which accommodates as many as possible
of the requirements. From both a management and an economic
point of view it is imperative that the majority of eggs are
laid in the nests provided. Floor eggs are difficult to collect
and soiled or broken eggs represent an economic loss. Eggs
that are laid on slats or on wire floors can be lost in seconds
by being trodden on. If sufficient are lost in this way, production will appear poor even if the hens are laying well.
Furthermore, broken floor eggs can lead to egg eating problems.

Work in this area will not be easy, past results having
produced few clues as to where to start. An observation study
by Perry et al. (1971) on a commercial flock which was laying
30% of its eggs on the floor during the 45th week in lay, could
find no relationship between the most favoured spots and any of
the variables, temperature, air speed or light intensity. Thus,
the physical parameters which can be easily measured may not be
important in determining the site that is selected.

CONCLUSIONS

Although there is much information available on the
behaviour of the laying hen, it is not always appropriate to
the commercial context. The ultimate aim of the welfare
pressure groups is to impose restrictions on commercial practice

and to provide a more 'humane' system of housing birds than
they view cages to be. At the present time, the scientific
evidence that is available does not permit precise regulations
to be formulated for floor housing systems. The most pressing
problems tend to be husbandry orientated rather than purely
scientific, although the latter have a role to play in establishing the causation of behaviour. For example, if it is
known why a bird chooses to lay in a particular spot, perhaps
a better nesting system can be developed.

In many of the reviewed papers large strain differences
were reported. Some of the researchers used birds that are no
longer commercially available and several centres have maintained their own breeding populations for many years. Any work
which is designed to assess the suitability of a particular
husbandry system should incorporate modern commercial stocks.
It may even be possible to breed birds that are particularly
suited to one environment.

REFERENCES

Bailey, B.B., Quisenberry, J.H. and Taylor, J., 1959. A comparison of
    performance of layers in cage and floor housing. Poult. Sci. 38:
    565-568.

Bareham, J.R., 1972. Effects of cages and semi-intensive deep litter pens
    on the behaviour, adrenal response and production in two strains of
    laying hens. Br. Vet. J. 128: 153-163.

Biswas, D.K. and Craig, J.V., 1971. Social tension indexes and egg
    production traits in chickens. Poult. Sci. 50(4), 1063-1065.

Christmas, R.B., O'Steen, A.W., Douglas, C.R., Kalch, L.W. and Harms, R.H.,
    1974. A study of strain interaction of cage versus floor layers
    for three evaluation periods at the Florida Evaluation Centre.
    Poult. Sci. 53: 102-108.

Craig, J.V. and Bhagwat, A.L., 1974. Agonistic and mating behaviour of
    adult chickens modified by social and physical environments. Appl.
    Anim. Ethol., 1: 57-65.

Craig, J.V. and Guhl, A.M., 1969. Territorial behaviour and social
    interactions of pullets kept in large flocks. Poult. Sci. 48: 1622.

Dorminey, R.W., 1974. Incidence of floor eggs as influenced by time of
    nest installation, artificial lighting and nest location.
    Poult. Sci., 53: 1886-1891.

Duncan, I.J.H. and Wood-Gush, D.G.M., 1972. Thwarting of feeding
    behaviour in the domestic fowl. Anim. Behav. 20: 444-451.

Gowe, R.S., 1955. A comparison of the egg production of seven white
    Leghorn strains housed in two environments - floor pens and laying
    battery. Poult. Sci., 34: 1198.

Guhl, A.M., 1953. Social behaviour of the domestic fowl. Kans. Expt.
    Stn. Tech. Bull. 73.

Hughes, B.O., 1975. The concept of an optimum stocking density and its
    selection for egg production. In: Economic factors affecting egg
    production. 271-298. Eds. B.M. Freeman and K.N. Boorman. Brit.
    Poult. Sci. Ltd., Edinburgh.

Hughes, B.O., 1977a. Some implications of dominance hierarchies in intensive
    husbandry systems. Appl. Anim. Ethol. 3(2), 199.

Hughes, B.O., 1977b. The absence of a relationship between egg production
    and dominance in caged layer laying hens. Br. Poult. Sci., 18:
    611-616.

Hughes, B.O. and Black, A.J., 1976. Battery cage shape: its effect on diurnal feeding pattern, egg shell cracking and feather pecking. Br. Poult. Sci., 17: 327-336.

Hughes, B.O. and Black, A.J., 1978. Agonistic behaviour in domestic fowls transferred between cages and pens. Appl. Anim. Ethol., 4: 181-186.

Hughes, B.O., Wood-Gush, D.G.M. and Morley Jones, R., 1974. Spatial organisation in flocks of domestic fowls. Anim. Behav. 22: 438-445.

Hurnik, J.F., Jerome, F.N. and Reinhart, B.S., 1971. The effect of colour and position on the choice of nesting location by the domestic hen. Poult. Sci., 50(5), 1587.

Hurnik, J.F., Jerome, F.N., Reinhart, B.S. and Summers, J.D., 1973. Colour is a stimulus for the choice of nesting site by laying hens. Br. Poult. Sci., 14(1), 1-8.

Hurnik, J.F., Reinhart, B.S. and Hurnik, G.I., 1973. The effect of coloured nests on the frequency of floor eggs. Poult. Sci., 52(1), 389-391.

James, J.W. and Foenander, F., 1961. Social behaviour studies on domestic animals. 1. Hens in laying cages. Aust. J. Agric. Res., 12: 1239-1252.

Johnson, E.A., 1964. Seven years of floor and cage egg production at the California Random Sample Test. Rep. 14th Calif. Off. Random Sample Egg Laying Test. 1962-1963.

Lowry, D.C., Lerner, I.M. and Taylor, L.W., 1956. Intra-flock genetic merit under floor and cage managements. Poult. Sci. 35: 1034-1043.

McBride, G., 1960. Poultry husbandry and the peck order. Br. Poult. Sci. 1(1), 65-68.

McBride, G., 1964. Social behaviour of domestic animals. II. Effect of the peck order on poultry productivity. Anim. Prod. 6(1), 1-7.

McBride, G., 1970. The social control of behaviour in fowls. In: Aspects of Poultry Behaviour. Eds. B.M. Freeman and R.F. Gordon. Brit. Poult. Sci. Ltd., Edinburgh.

McBride, G. and Foenander, F., 1962. Territorial behaviour in the domestic hen. Nature, London, 194: 102.

Miller, M.M., 1956. Factors affecting egg production, body weight and feed efficiency of selected strains of cage layers. Texas A & M College M.Sc. Thesis.

Perry, G.C., Charles, D.R., Day, P.J., Hartland, J.R. and Spencer, P., 1971. Egg laying behaviour in a broiler parent flock. WPSA Journal 27(2), 162.

Polley, C.R., Craig, J.V. and Bhagwat, A.L., 1974. Crowding and agonistic behaviour: A curvilinear relationship? Poult. Sci., 53: 1621-1623.

Savory, C.J., 1980. Diurnal feeding patterns in domestic fowls: a review. Appl. Anim. Ethol. 6(1), 71-82.

Shupe, W.D. and Quisenberry, J.H., 1961. Effect of certain rearing and laying house environments on performance of incross egg production type pullets. Poult. Sci. 40: 1165-1171.

Tindell, D. and Craig, J.V., 1959. Effects of social competition on laying house performance in the chicken. Poult. Sci. 38: 95-105.

Wood-Gush, D.G.M., 1972. Strain differences in response to sub-optimal stimuli in the fowl. Anim. Behav. 20: 72-76.

Wood-Gush, D.G.M., 1975. Nest construction by the domestic hen: some comparative and physiological considerations. In: Neural and endocrine aspects of behaviour in birds. Eds. Peter Wright, Peter, G. Caryl and David M. Vowkes. Elsevier Scientific Publishing Co., Amsterdam.

Wood-Gush, D.G.M. and Gilbert, A.B., 1969. Oestrogen and the pre-laying behaviour of the domestic hen. Anim. Behav., 17: 586-589.

Wood-Gush, D.G.M. and Murphy, L.B., 1970. Some factors affecting the choice of nests by the hen. Brit. Poult. Sci., 11: 415-417.

DISCUSSION

J. Fris Jensen *(Denmark)*

Your comment on the round food trough suggested that it may have some advantage compared with the single row. We have some observations that when you put hens on a round food trough it leads to a higher pecking frequency because of the position they are standing in. The birds peck at their neighbours.

J.A. Hill *(UK)*

Yes, but of course this also happens on ordinary long troughs on cages. You do get birds that will peck birds in the adjacent cages, and sometimes it can be quite fierce. There are pros and cons, I am just suggesting some of the arguments that could be put forward and saying that I think it is very difficult to make firm decisions about things like trough space allocation because with different feeders you may have a totally different requirement.

J.M. Faure *(France)*

You said that when comparing production results in pens and battery cages, there is a lot of discrepancy. I do not find this surprising. I have looked at the results of annual tests, comparing about ten commercial strains, all at nearly the same level of production. They are tested in cages and in pens. What is very clear is that if you take the results from ten years ago, when virtually no strains were selected for cages, all the results are better in pens than in cages. If you take the most recent results, where nearly all the strains have been selected for cages, then you see that all the results are better in cages than in pens. This shows that at least as far as production is concerned, it is possible to select birds which produce better in cages than in pens and it explains the discrepancies you mentioned.

J.A. Hill

Yes, that is possibly a large factor.

R. Tauson *(Sweden)*

    I can't remember what your manure removal system was.
Did you have a common pit for both tiers?

J.A. Hill

    Yes, that's right.

R. Tauson

    So that means that the manure from the top tier fell
through the net down to the first tier. Did you not have any
hygiene problems with the feed and water?

J.A. Hill

    No, there were manure deflectors for the feeders and
drinkers so it was impossible for the manure to fall into a
feeder or a drinker. The other question of course is dirty
birds. As I pointed out, we terminated the experiment after
five months and the birds had not become particularly dirty in
that system. They do tend to move about quite a lot and not
sit under the slats where other birds are dropping things on
top of them. In this respect the situation was much better
than in the get away cage where we did get atrociously dirty
birds because they were always confined underneath other birds.

W. Fölsch *(Switzerland)*

    I would like to stress that the method of rearing is
very important in relation to the number of eggs you will
finally find in cages or deep litter. We had hens reared in
deep litter and put in cages, and vice versa, hens reared in
cages and then put onto deep litter for the production period.
The birds that were reared in cages and then put onto deep
litter did not use the nests; they went into a corner all
together, laid the eggs and the eggs were destroyed if we did
not pick them up immediately. So we collected eggs every hour
in all our systems for three days. We had the same number of
eggs then from all systems. In the periods before and after

the hourly collection we found we had less eggs in those systems where the birds had been transferred from one system to the other for the production period.

J.A. Hill

Yes, I would not disagree with that. I think the rearing is important. In fact, the birds in this slat system were reared in cages, which may have had a bearing on it. However, having said that, my experience with broiler/breeder flocks which are nearly always kept on litter in our country, and are reared on litter as well, shows that there is tremendous variation in whether or not they use nests which are provided for them. It can be a variation not only between strains, but from flock to flock in the same house where the nests, the environment and everything else are almost identical. So, I think there is a lot we don't know yet.

K. Vestergaard (Denmark)

I have a comment on egg eating. In our deep litter house we made a video tape recording and it showed a lot of egg eating and birds damaging the eggs, so there were lower production figures. It is significant that in dealing with welfare, production figures are sometimes quoted as an indication. It should be remembered that the cage always has the advantage over other systems in that the percentage of eggs which are damaged or destroyed between laying and collection is much lower.

I have another comment about the hens feeding together in the shallow cages and the narrow cages which you mentioned. It is clear if you look at the social stimulation aspect that it must be much stronger in the cage where the birds are very close to the feeding area. Birds are much more likely to feed together than in the deep litter system where many birds are not close to the feeding area.

J.A. Hill

I think there is quite a lot of evidence to suggest that
the reason for the particular pattern of feeding that you get
in laying birds is not, in fact, an effect of social facilitation
or that they are all in a cage grouped together, but is due to
a physiological requirement for egg formation and so on at
certain times in the ovulatory cycle. It probably has a physiological basis rather than a behavioural one in this particular
case; the pattern is very marked.

R-M. Wegner (FRG)

First of all a comment about dirty birds. We have had
no problems with dirty birds in the 'get away' and aviary
systems. In the 'get away' cages, if you fix the perches in
the right position there is no difficulty.

Now I have two questions. The ethologists are always
saying that it is necessary for all the birds in the same cage
to be able to feed at the same time. I do not quite understand
why that is so essential. I remember the first battery cages
developed in England; they were designed in such a way that the
birds had only sixty minutes in the day to take up feed but it
was quite long enough to get a very good performance. Today
the troughs are in front of the cages all the time and the
laying hen has 14 or 16 hours per day to take up the feed. Why
is it so essential that all the laying hens should be able to
feed at the same time?

My second question is also to the ethologists and I hope
I will not get too many different answers. Is a dust bath
really essential for the 'get away' cage, and do you think the
'get away' cage will be the solution to the cage problem?

J.A. Hill

With regard to the question of the birds feeding at the
same time, I am not proposing that they necessarily should be
able to feed all together - that is not what I said. I said

224

that the question could be posed because there is evidence
that if birds are thwarted in feeding behaviour they show
symptoms of frustration.

I.J.H. Duncan *(UK)*

On the same point, there is a tendency for birds to
synchronise their activities and there could be two possible
reasons for this. One is that common factors are having the
same effects on all the birds, for example, we know that they
show a diurnal rhythm in feeding and that there does seem to be
a tendency to feed towards the end of the day even if the birds
are kept individually. The second reason why they may synchronise
their activities is that they are a social species and tend to
do the same things at the same time.

B.O. Hughes *(UK)*

Cage design is an additional factor. In our cages
which have vertical fronts, it is often not possible for more
than two birds at a time to feed because of the way they put
their heads through the bars, even though the width is 40 cm.
In theory, at least three birds should be able to feed in that
width. There may be a third bird which wants to feed in which
case there will be a lot of jostling for position. Obviously,
if this could be avoided it would be desirable.

G. Martin *(FRG)*

I agree; I have observed that hens which do not have
enough space push each other to get to the trough. They will
push under and over each other to get to the food.

I.J.H. Duncan

Although it may seem sensible to us to queue up for our
dinner I do not think it is reasonable to expect the lower
animals to do this.

R-M. Wegner
_____

Are there any figures available on how many birds will be frustrated if they cannot get to the feed troughs at the same time?

B.O. Hughes
_____

I have some data on feeding patterns and numbers of birds feeding at any one time. The results don't actually show how many birds were frustrated but do give an indication of variation in terms of the numbers feeding - two feeding, two not, or three feeding, one not, etc - at different times of the day. So there is some indication but not a quantitative indication.

K. Vestergaard
_____

It seems that it is still necessary to include dust baths in our experiments. With regard to feeding behaviour, it seems that feeding may also elicit aggression so perhaps we should reconsider the cafeteria system where the birds can feed adequately in a very short time. It is odd that the birds do feed for so long a time from a trough; it may be that they peck the food for other reasons. Perhaps a lot of the pecking is, in fact, bill raking - the first element in dust bathing.

THE PRE-LAYING BEHAVIOUR OF LAYING HENS IN CAGES WITH AND WITHOUT LAYING NESTS

G.C. Brantas

The nesting or pre-laying behaviour in a hen house with laying nests has been extensively studied, principally by Wood-Gush (1963, 1971) at the Poultry Research Centre in Edinburgh. I will give a general description of the complete pre-laying behaviour in a hen house with laying nests although not every detail is always seen.

Phase 1

The bird is restless and searches for a place to lay an egg. Some birds omit this phase and go directly to the nests.

Phase 2

The hen inspects a number of nests between other activities such as feeding, preening, sleeping. The nest inspections are intention movements. Gradually the hen moves most of its body into the nest; finally she enters the nest.

Phase 3

The hen lies down, makes a small hole in the nesting material and picks at this. She does not move her position. She lays an egg.

Phase 4

After laying the egg, the hen remains sitting and examines the egg, visually or using her beak. She then rises and leaves the nest, cackling, to join the flock.

In a battery cage very different laying behaviour may be observed. The following description I derive from Martin (1975). The searching behaviour in the cage consists of search

R. Moss (ed.), The Laying Hen and its Environment, 227-237

movements and the pushing away of other hens in the cage. The
neck protrudes as the bird searches the cage floor and its
corner, emitting weak sounds. The search movements increase in
intensity. Sounds of fright are heard. The hen creeps between
the legs of another hen to seek cover. One hen observed
repeated this pattern more than 100 times. Then the hen appears
to make an attempt to escape; she puts her head between the
wires of the cage and squeezes her body forward. Although she
is then almost always picked at by the birds in neigbouring
cages, she continues. She tries to fly; the tail is extended
and the plumage is smoothed; the head moves up and down with
increasing speed; the wings are slightly raised; the body is
elongated. The hen sometimes attempts to climb the cage and
falls down. The behaviour can give an appearance of panic.
Then quite suddenly the hen no longer tries to escape and she
sits down quietly and lays her egg. The hen is quite exhausted:
her breathing is faster, her sound weakened, and she sometimes
sinks down. Though the descriptive difference in the behaviour,
is very great one can also provide quantitative data. Martin
(1975) has given some figures. In her observations the whole
egg laying procedure lasted on average 16.4 minutes per hen in
a hen house with laying nests. Transfer of the birds to two-
hen cages increased the time of pre-laying behaviour to 74.2
minutes. If, thereafter, the hens were placed singly in cages,
the time was reduced to 51.4 minutes. When returned to the hen
house with laying nests the time fell to 16.5 minutes.

	Hen house with laying nests	2-hen cage	1-hen cage	Hen house with laying nests
Pre-laying time:	16.4 min.	74.2	51.4	16.5 min.

One can criticise these figures in two respects. First,
the environments studied, hen house, 2-hen cage and 1-hen cage
differ from each other in many ways: laying nest or not, nesting
material or not, more or less space, possibility of escape or

not, different social environments, and so on. Therefore one
cannot know what is the factor causing the difference in
behaviour. The second criticism is more important; pre-laying
behaviour starts with an increase of search movements and with
some disquiet. It does not start suddenly, but it steals in
slowly and can pass unperceived. Consequently a slight error
in observation of the height of the disquiet of the hen will
result in considerable error in fixing the time of the beginning
of the disquietness, and therefore with the beginning of the
pre-laying behaviour.

In order to study pre-laying behaviour and avoid the
drawbacks, Martin, two students and I carried out an investigation into cages with a laying nest attached. The cages
measured 50 x 45 cm and each one gave access to a wooden
laying-nest of 30 x 40 cm with wood shavings in it. It was
possible to close the access to the laying nest. Thus there
was only one factor to study: cages with and cages without
laying nests.

Furthermore, I did not measure the duration of disquietness,
but figures of other parameters of disquiet during five-minute-
periods.

When the animal had laid her egg, and the time of egg
laying was known, it was then possible to reconstruct the
figures of the parameters for the last quarter of an hour
before, as well as the last half-an-hour before and the last
hour before egg laying.

As indications of disquiet I used the following parameters:

The number of paces the animal makes. Pacing is stereotyped back and forward movement. Duncan (1970) has
proved that pacing occurs when the level of frustration
is higher. At first in a frustrating situation the pacing
movements look like escape movements, but later they

become stereotyped. The existence of stereotyped pacing is indicative of the fact that the animal is frustrated or has been frustrated.

The number of creepings under another animal to seek cover.

The number of attempts to escape.

The number of attempts to sit down without sitting for more than 3 seconds.

The number of real sitting periods of more than 3 seconds.

The average duration of a sitting period.

The duration of the total sitting time.

From the viewpoint of disquiet I suppose the inter-pretation of these parameters is obvious.

I carried out my investigation with 98 animals; 68 White Leghorns, Hisex and Shaver and 32 brown animals, Derco. They were divided into two groups. One group started in cages with access to the laying nest. After 3 weeks the access to the laying nest was closed; after 3 weeks the access was restored. This first group was the access-closed-group. The second group was the closed-access-group.

Now I will give you some data. I will restrict myself to a representative selection. All of these data concern the White Leghorns; figures in relation to the brown birds were similar (see Tables 1 - 4).

The cages that have been studied had 1, 2 or 4 birds in them. The differences in the parameters of disquiet during the pre-laying period were very small in these more or less crowded cages and therefore I will only make one remark and will not give you further data.

TABLE 1

Access-closed-access group, last half an hour before egg laying			
	Access	No access	Access
Paces	53	220	64
Creepings	0.0	2.2	0.0
Attempts to escape	0.0	0.9	0.0
Attempts to sit down	0.8	3.4	1.6
Sitting periods	7.6	6.6	7.5
Duration sitting period	3.5	1.0	3.4
Sitting time	19.0	1.0	3.4

TABLE 2

Closed-access-closed group, last half an hour before egg laying			
	No access	Access	No access
Paces	275	76	419
Creepings	1.1	0.0	1.5
Attempts to escape	0.2	0.0	1.7
Attempts to sit down	6.6	1.2	3.5
Sitting periods	5.9	8.9	6.4
Duration sitting period	0.5	2.8	0.4
Sitting time	3.4	18.6	3.9

TABLE 3

Last hour before egg laying		
	Access of laying nest	No access
Paces	105	434**
Creepings	0.0	0.2
Attempts to escape	0.0	0.3
Attempts to sit down	2.0	8.9*
Sitting periods	11.7	11.4
During sitting period	3.6	0.5**
Sitting time	34.4	5.3**

TABLE 4

Last quarter of an hour before egg laying		
	Access to laying nest	No access
Paces	25.0	141.0**
Creepings	0.0	0.9
Attempts to escape	0.0	0.1
Attempts to sit down	0.5	3.6*
Sitting period	3.9	3.7
Duration sitting period in min.	4.1	0.6**
Sitting time	10.1	2.2**

*Significant on 0.05 level
**Significant on 0.01 level.

In the 3 weeks during which birds stay in a cage in a special situation (access or no-access to the laying nest) they adapt themselves somewhat when there is no access to the laying nest. This adaptation showed only in the 'pacing' parameter with any statistical significance (P = 0.05). However, in a no-access situation after 2 weeks of adaptation the number of paces during the last half an hour before oviposition was still more than 4 times greater than in the access-to-nest situation after 2 weeks of adaptation; thus I suggest that the birds will never adapt themselves completely to a no-access-to-laying nest-situation.

What conclusion do we reach?

In my opinion it is quite clear that without a laying nest the pre-laying disquiet is very much greater than in a situation with a laying nest. I firmly believe this even if physiological measurements do not support this conclusion by not showing any difference between the nest and no-nest situations. I stress the fact that physiology has not given contradictory information.

My quantitative findings are a confirmation of a part of
the qualitative description of the pre-laying behaviour in
cages as made by Martin (1975).

In my opinion disquiet is an indication of frustration,
of lack of welfare. The disquiet caused by the lack of a laying
nest is so great that in my opinion and for this reason only
the battery cage must be thought as very inadequate to fulfil
the behavioural needs of the laying hen.

REFERENCES

Duncan, I.J.H., 1970. Frustration in the fowl. In: Aspects of poultry
     behaviour. Edit Freeman, B.M. and Gordon, R.F., Br. Egg Market
     Bd. Symp. No.6 pt 2. Br. Poult. Sci., Edinburgh.

Martin, G., 1975. Ueber Verhaltensstörungen von Legehennen im Käfig. Ein
     Beitrag zur Klärung des Problems Herschutzgerechter Hühnerhaltung.
     Angew. Ornithl. 4, 145-176.

Wood-Gush, D.G.M., 1963. The control of the nesting behaviour of the
     domestic hen. Anim. Behav. 11, 2/3, 293-299.

Wood-Gush, D.G.M., 1971. The behaviour of the domestic fowl. London,
     Heinemann, ISBN O-435-62920.4.

DISCUSSION

I.J.H. Duncan *(UK)*

Can you tell us what was the experience of your hens before they were put in the experimental situation?

C.C. Brantas *(The Netherlands)*

They came from a deep litter situation, with nest boxes.

C. Beuving *(The Netherlands)*

As I was saying yesterday, we found an increase in corticosterone secretion during egg laying which must be ascribed to egg laying behaviour and not to disturbed egg laying behaviour. I have no results on the additive effect when two stresses are applied together. Perhaps that is one possibility if there is no additive effect. Another possibility is that stress can perhaps be defined by other measures than corticosterone production.

C.C. Brantas

We must accept that it is possible that ethology can establish some things which are not confirmed by physiology. As long as physiology does not prove the contrary then ethology has the right to speak.

W. Bessei *(FRG)*

I would like to pursue the point raised by Dr. Duncan. Don't you think that the experience of nest boxes which the hens had before the experiment would lead to a higher frustration situation? In my opinion a closed nest box would lead to more frustration than no nest box at all.

C.C. Brantas

I agree that the frustration of being without a nest box is greater for birds which have been used to one than for birds which have never had one. However, I do not think the difference is sufficient to revise these figures.

I.J.H. Duncan

May I ask why you used the description of Wood-Gush for nesting in pens and the description of Martin for nesting in cages? That is only adding variability. Wood-Gush has described the behaviour of hens nesting in cages. Would that not have made the comparison a little more valid?

C.C. Brantas

I did it because Dr. Martin's description was more emotional and more subjective. However, that does not make it untrue. Even such an emotional description can be confirmed with figures.

J. Fris Jensen *(Denmark)*

You presented the total movements over the three week period. Have you any information on the variation during the three week period?

C.C. Brantas

No. It was difficult enough to collect the data we have. My students started to observe the animals without knowing if and when an egg would be laid. One student watched a bird for eight hours; he took a quarter of an hour off for a cup of coffee and in that time the egg dropped!

W. Bessei

You mentioned briefly that you had the impression that there were differences between the breeds. You have only presented results for the White Leghorn. What about the others?

C.C. Brantas

I do not know; I only observed three strains - two white ones and one brown one. There is not enough time in my life to observe all the strains that exist.

W. Bessei

I know, but you have only presented results with the White Leghorn.

C.C. Brantas

The results with the brown animals were essentially the same.

W.F. Raymond

Thank you. There will be time for general discussion later but we must move on now to the next paper.

SOME SYSTEM DEFINITIONS AND CHARACTERISTICS

C.M. Hann

I will introduce two subjects in this talk; the first, a basic one - that of definitions and terminology. This will lead neatly to the second, which is an attempt to summarise some of the characteristics of systems.

DEFINITIONS

The forces which have shaped today's poultry industry stretch back many decades - some into the past century. Thus it is not surprising that much is taken for granted when we talk about poultry production. However, in our current role I believe it is essential we should try to establish a common ground so that misunderstandings can be minimised. In Table 1 brief descriptions of a number of systems have been attempted. In some cases I hope there is little room for argument, although it may well be that there can be some refinement; in others, where I have been rather arbitrary, discussion may help us to arrive at mutually acceptable wording.

Some issues, such as properly defining free range, have been taxing legal minds for some years without producing an acceptable definition. It may be easier to decide what is not free range, rather than what is.

A less difficult matter is to agree what may be regarded as a 'colony'. At what point does a multi-bird cage become a colony cage?

While I suggest we could spend a short while discussing some of these points, the issues raised by Table 2 are likely to be more controversial.

R. Moss (ed.), The Laying Hen and its Environment, 239-258

A GUIDE TO EGG PRODUCTION SYSTEMS

SYSTEM	DESCRIPTION
Cages	Typically a cage laying system consists of a series or battery (hence the term laying battery) of cages each designed to hold one or a number of birds. The cages have sloping mesh floors so that the eggs, when laid, roll forward out of reach of the birds to await collection. Droppings pass through the mesh floors on to boards, belts, the floor of the house or into a pit to await removal. Feed is supplied in troughs fitted to the cage fronts and an automatic water supply is provided.
	Most cages are constructed of either wood and wire or metal. Plastic cages are also used to a limited extent. Wood and wire cages have a timber framework with 25 mm diameter mesh wire netting or 25 mm square mesh welded wire stretched on frames to form the cage floors and divisions. Metal cages have steel frames and usually 50 mm x 25 mm mesh welded wire floors. Cage divisions may be of solid sheet material or of wire construction. The wire and other metal parts are galvanised after fabrication.
Colony cages	It is suggested that this term be used for cages containing 8 or more birds.
Floor systems	Characterised by greater freedom of movement than in cages. Arising from the greater freedom the arrange- ments for meeting the requirements of the stock are rather different. Feeders and drinkers (waterers) are not integral (as with cages) and nesting accommodation has to be provided.
Deep litter	A flock system in which the stock is confined in a building with access to an area of litter material such as wood shavings, straw, peat moss, etc.
Wire or slatted floors	A flock system in which the stock is confined to an area of slats or wire mesh, or combinations of slats and wire, with or without perches.

FORMS	COMMENTS
From flat deck (single tier) to multi-tiers up to 6 or more high in vertical, stepped or semi-stepped configurations.	The cage system favours the application of mechanisation to the various routine activities.
Although density of stocking is commonly between 400 and 500 cm^2 per bird, it varies from under 400 cm^2 to more than double this amount, being influenced by such factors as bird body size, numbers of birds per cage, ambient temperature and any regulations that may exist.	The system enables house stocking densities to be achieved that facilitate the maintenance of optimum environmental temperature in cool climates. The adequate bird body heat is conserved by controlled ventilation and the insulation of the house structure.
Various forms. Stocking densities are similar to those for ordinary cages, but more space tends to be provided for breeding stock.	Some minor economies in cage construction (fewer partitions). In general, performance levels per bird are less good than with small groups. Of value for natural mating breeding units, e.g. 1 male and about 12 females or 3 males and 40 females.
As below.	A direct consequence of the greater freedom provided for the birds is the greater opportunity for problems to occur, particularly where large numbers of birds are involved. There is general agreement that to safeguard bird welfare as well as to achieve satisfactory performance, higher standards of husbandry and management are needed than for cage systems.
The main variable is the proportion of space that may be occupied by a platform. The range is from no platform at all (when moveable perches may be used) up to about 90% platform and only 10% litter. A common arrangement is 50% platform. Stocking densities range from about 750 cm^2 up to about 2000 cm^2 per bird according to the presence or absence of a slatted or wire mesh platform and to bird type and size.	Widely used for breeding flocks at primary breeding and multiplication (parent and grandparent) stages.
The floors may be flat or sloping to assist egg collection - as in that known as the Pennsylvania or Bressler system. In both cases nesting accommodation is necessary. Stocking densities range typically between 600 and 800 cm^2 per bird.	Used fairly widely in Denmark. Also to some extent for naturally mating breeding flocks.

A GUIDE TO EGG PRODUCTION SYSTEMS (Continued)

SYSTEM	DESCRIPTION
Aviary	The essence of this system is the exploitation of vertical space (as can be achieved with tiered cages) while allowing the birds to live as a flock. The main features are:
	(a) slatted and/or wire mesh platforms at more than one level carrying feeders and drinkers.
	(b) inter-connecting ladders;
	(c) easily accessible nest accommodation;
	(d) there may be a littered area at one of the levels.
	By using much of the volume of a building in this way, densities in terms of usable floor can be kept at around 900 cm^2 per bird while the use of the ground level area of the building can be as high as about 500 cm^2 per bird.
Straw yard, also pole barns and covered yards	The straw yard system normally combines an enclosed housing area adequate for roosting in at night (and containing feeders and nest boxes) with a usually more extensive yard area less than fully protected from the weather.
	Pole barn or covered yard accommodation comprises a covered area without full protection from the wind. Adjustable screens may give protection in bad weather.
	In all cases a liberal use of straw or other suitable litter is a fundamental feature necessary to ensure acceptable conditions for the stock.
Verandahs	This system consists of an enclosed roosting compartment with access to a larger area more or less open to the air. The whole of the accommodation is slatted and/or wire floored and so arranged that there is space for the droppings to accumulate underneath.
Semi-intensive (a) Static	The semi-intensive system combines housing for the birds, normally at densities of between 1500 and 2000 cm^2 per bird, with more or less regular day time access to an area of ground (grassland, soil, etc.) enclosed by fencing so that the stock is at a density in excess of 600 birds per ha.
(b) Fold units	This form of semi-intensive housing involves placing small groups of birds (usually less than 100 birds, commonly about 25 birds) at very high densities on successive patches of ground by moving the units every one or two days.

FORMS	COMMENTS
Many configurations are possible. Some trial results suggest that the system may function better when a littered area is included.	Limited trials with broiler breeder flocks have given promising results.
	Due to the extra platforms, ladders, etc. some difficulties in servicing aviary flocks may be unavoidable.
	This is probably the only flock system in which stocking density can be sufficiently high for there to be adequate bird body heat produced to enable the house temperature to be maintained close to the optimum in cool climates.
Often making use of existing farm buildings and/or improvised housing, straw yards vary considerably. The range extends from fully covered and enclosed yarding (which might give rise to conditions difficult to distinguish from deep litter) to fully open yards offering little weather protection for the stock.	Because such wide variation exists in terms of protection from the weather, it is particularly necessary to include an adequate description when referring to this system.
The minimum protected accommodation would normally be about 700 cm^2 per bird, while a typical yard area would be 2000 or 3000 cm^2 per bird.	The requirement for large quantities of suitable litter material limits the system to those areas where these are cheaply available.
Often constructed by placing fold units on slatted or wire mesh platforms, types range from fully roofed to units with only the roosting compartment roofed. Units normally quite small, eg, 30 to 200 birds. Density range 400 to 800 cm^2 per bird.	Appears to have few advantages.
Housing is usually semi-permanent and may be slatted or solid floored, the latter being more common. A popular arrangement is for access to two separate pens to be provided from each house. By this means one patch of ground can be allowed to recover whilst the other is in use.	The carrying capacity of the land depends on soil type (on light soils 700 or 800 birds per ha may be satisfactory, on heavy soils only about 600) and on good management. Housing should be such that the birds can be kept inside during bad weather.
Various forms, all but the smallest units being moved by tractor. Density on the unit commonly between 1000 and 1500 cm^2 per bird. Sleeping accommodation may be slatted or perches and no floor.	System only really suitable for well drained land that is fairly flat.

A GUIDE TO EGG PRODUCTION SYSTEMS (Continued)

SYSTEM	DESCRIPTION
Free range	Moveable, semi-fixed or fixed housing with more or less regular day time access to the ground at stocking densities of not more than 300 birds per ha on a continuous occupancy basis. Higher stocking densities are permissible for limited periods.
	The name implies freedom for the birds to roam widely and unconstrained by fences, however, the provision of effective perimeter fences helps to deter predators.

FORMS	COMMENTS
The main variable is the type of house which, if moveable, is fairly small eg, 3 m x 2.5 m and may be slatted floored, the stocking rate being around 500 or 600 cm^2 per bird. Fixed housing may be larger, eg, 7.5 m x 4 m and is often solid floored with space allowances up to 2000 cm^2 or so per bird. Various devices are used to reduce damage to the ground around the houses.	Combinations of free range and semi-intensive or other systems exist in which the free range occurs during better weather and the stock is more closely confined at other times, eg, in the winter. In view of the temptation for some producers to claim free range egg production when this is unjustified, clear guidelines are likely to be of value.

CHARACTERISTICS

Any consideration of the welfare of livestock under alternative systems has to take into account the fact that there is no perfect system and that the advantages and disadvantages of one have to be compared with those of another. Anyone who says system B is to be preferred to system A is really saying that in their estimation the balance of advantages lies with system B. In Table 2 I have tried to look at some, and only some, of the factors that ought to be examined. I would warn you that the method I have used is very approximate and arbitrary and a number of the assessments and interpretations are highly subjective.

However, I still feel the approach is a sensible one because it emphasises that we are not just talking about systems, but also the ways in which those systems will be operated.

The cage system has become widespread and popular because it is less demanding on those who look after the stock. Almost anyone with reasonable common sense can operate a cage unit and get good or very good results. There is little to go wrong and when troubles do develop the solutions are often fairly easy.

By contrast, the alternative systems which have been largely abandoned over the past 15 years, tend to require more labour and present greater challenges to those who operate them. Most people aim for an easier life with fewer problems - hence the drift towards cages, even at times when cages have not been a cheaper form of housing.

The consequences of indifferent or poor management and husbandry are rarely adequately considered. In our current discussions it is vital that the qualities and limitations of the farmers and farmworkers concerned are taken into account. I should explain at this point that I have taken 'husbandry'

AN ATTEMPT TO ASSESS SOME PRODUCTION CHARACTERISTICS AND CIRCUMSTANCES ACCORDING TO SYSTEM AND LEVEL OF SKILL EMPLOYED   Part 1

The reference standard used for comparison in each case is that appropriate to cage production under average husbandry/management. The data in this Table are illustrative and not definitive.

System	Level of husbandry/ management (1)	OUTPUTS			INPUTS				
		Eggs collected (2)	Dirty eggs (3)	Survivors (4)	Feed consumed (5)	Labour (6)	Land (7)	Capital (8)	Medication and additives (9)
Cages	Good	○○	○○	○○	○○	○	○	○	○○
	Average	○	○	○	○				○
	Poor	●	●	●	●				●
Litter	Good	○○	●	○	●●				○
	Average	●	●●	●●	●●●	x1.2	x2.0	x1.5	●●
	Poor	●●●	●●●	●●	●●●				●●●
Wire floor	Good	○	●	○	○				○○
	Average	●	●●	●	●●	x1.2	x2.0	x1.2	○
	Poor	●●	●●●	●●	●●				●
Aviary	Good	○	●	○	○				○○
	Average	●	●●	●	○	x1.3	x1.0	x1.1	○
	Poor	●●	●●●	●●	●●				●
Straw yard	Good	●	●	●	●●				●
	Average	●●	●●	●●	●●●	x2.5	x3.0	x1.4	●●
	Poor	●●●	●●●	●●●	●●●●				●●●
Verandahs	Good	●	●	●	●				○○
	Average	●●	●●	●●	●●	x2.0	x2.5	x1.5	○
	Poor	●●●	●●●	●●●	●●●				●
Semi-intensive	Good	●	●●	●	●				●
	Average	●●	●●●	●●	●●	x6.0	x80	x1.8	●●
	Poor	●●●	●●●●	●●●	●●●				●●●
Free range	Good	●	●●	●	●				●
	Average	●●	●●●	●●	●●	x10	x330	x1.4	●●
	Poor	●●●	●●●●	●●●	●●●				●●●

Part 2

System	Level of husbandry/ management (1)	ENVIRONMENTAL FACTORS						
		Temperature maintenance (10)	Dampness and wetness (11)	Dust levels (12)	Int. parasites & enteric D. (13)	Ext. parasites (14)	Predation (15)	Rodents (16)
Cages	Good	○○	○	○	○	○	○	○○
	Average	○	○	○	○	○	○	○●
	Poor	●	●	●	○	●	○	●●
Litter	Good	●	○	●●	●	○	○	○
	Average	●●	●	●●	●●	●●	○	●●
	Poor	●●●	●●	●●●●	●●●	●●	○	●●●
Wire floor	Good	○	○	○	○	○	○	○
	Average	●	○	●	○	○	○	○
	Poor	●●	●	●●	●	●	○	●
Aviary	Good	○○	○	○○	○	○	○	○
	Average	○	○	○	●	○	○	○
	Poor	●	●	●●	●●	●	○	●
Straw yard	Good	●	○	○○	●	●	○	●●
	Average	●●	●●	○	●●	●●	●	●●
	Poor	●●●	●●●	○	●●●	●●●	●●	●●●
Verandahs	Good	●●	○	○	○	○	○	○
	Average	●●	●	○	○	●	●	●
	Poor	●●●	●	○	●	●●	●●	●●
Semi-intensive	Good	●●	●●	○	●	●	●	●●
	Average	●●	●●●	○	●●	●●	●●	●●
	Poor	●●●	●●●	○	●●●	●●●	●●●	●●●
Free range	Good	●●	●●	○	●	●	●	●●
	Average	●●	●●●	○	●●	●●	●●	●●
	Poor	●●●	●●●●	○	●●●	●●●	●●●	●●●

LEGEND

(1) Level of husbandry/management: Good   /bonne   /gut   husbandry/culture/bildung
                                   Average/moyenne /durchschnittlich   "   "   "
                                   Poor   /mauvaise/schlecht   "   "   "

(2) Eggs collected: eggs laid minus eggs lost or broken before collection
(3) Dirty eggs: eggs in need of cleaning - contamination with dirt or micro-organisms.
(4) Survivors: end of lay birds available for sale.
(5) Feed consumed: feed taken by the birds plus food wasted (spilt or taken by rodents, etc.)
(6) Labour: amount of labour required per bird.
(7) Land: amount of land required per bird.
(8) Capital: amount of capital required to set up new unit (1980), per bird.
(9) Medication and additives: likely requirement.
(10) Temperature maintenance: ability to maintain optimum temperature in cold weather.
(11) Dampness and wetness: likelihood of these occurring.
(12) Dust levels: likelihood of dusty conditions.
(13) Intestinal parasites and enteric diseases: likelihood of occurrence.
(14) External parasites: likelihood of occurrence and difficulty of control.
(15) Predation: risk of loss from foxes, birds of prey, etc.
(16) Rodents: risk of infestation - difficulty of control.

Progressively/progressivement more satisfactory/au meilleur/besser   oo ooo
       "   "   less   "   plus mauvais/schlechter   ● ●● ●●●

Standard - average husbandry/management in the cage system   o

CMH March, 1980

to mean the day-to-day process of providing the stock with its needs and recognising emerging problems. 'Management' is the organisation that makes it possible for a unit to function and the guidance and support necessary for those in direct contact with the stock. On many units these functions are vested in one man.

Using average levels of management and husbandry, combined with the cage system as a standard, Table 2 picks out the commercially vital input and output factors and tries to assess these in the light of both the systems and the degrees of skill used in the process of operating them.

I should emphasise that I do not for one moment consider the factors I have selected as being equivalent. For example, in commercial terms the value of eggs produced (column 2) is very much greater than that of the end of lay hen (column 4). Likewise in welfare terms, the benefits of a warm house (column 10) may not be comparable with, say, the presence or absence of parasites (columns 13 and 14). Further problems of comparison exist because changes do not all come about in the same way or may apply only in part of a system. For example, in summer time the free range that the birds enjoy is likely to be entirely dust-free (pollen particles excepted), but the inside of the range house may be more dusty than a deep litter unit.

In spite of such difficulties, I believe a methodical approach such as this is necessary before we can say that this or that system is likely to be better or worse when adopted by poultry farmers.

At this seminar there has been much discussion of a whole range of ethological factors. I have not tried to include any of these in my Table, mainly because adequate data are lacking. The programme of trials and experiments that we hope will emerge as a result of our deliberations may result in our being able to fill in many of the gaps.

DISCUSSION

W.F. Raymond *(UK)*

When I first saw Mr. Hann's Table it occurred to me that it would have been an interesting exercise had he distributed blank copies of it here and each of us had filled it in independently. I wonder if the collated results would have differed from those of Mr. Hann. However, this Table seems to me to be a most useful approach in bringing together the various aspects we have been talking about. Now, are there any specific comments?

R-M. Wegner *(FRG)*

In column No. 6, under the heading 'Labour' can you explain what the figure of 1.2 means?

C.M. Hann *(UK)*

It means you need 20% more labour than in an average cage system, i.e. 1.2 x the labour in an average cage system.

J.A. Hill *(UK)*

You stated early in your presentation that in cages there is very little to go wrong. With increased trends towards mechanisation of systems, mechanised feeding, egg collection and so on, I would disagree with you over that one. Certainly, people who look after cage systems nowadays have to be a little bit inclined towards mechanics and engineering.

C.M. Hann

That is probably fair comment. What was at the back of my mind was the historic development in which cages have become so dominant in most of Europe. In the United Kingdom 95%+ of commercial layers are in cages. Most of that trend occurred when mechanisation was limited to some manure disposal and a very limited amount of feed distribution. I agree that today two types of people are needed in cage units, a resident

engineer (if not an electronics engineer) as well as the
poultryman. However, the things that can go wrong are usually
linked to alarm devices so that appropriate action can be taken
if necessary. I am thinking more of the interrelationship
between the man and the bird, than of the man and the equipment
related to the bird, but I know this is a mix which is difficult
to segregate. You have made a good point.

I.J.H. Duncan *(UK)*

One common feature of most of the welfare codes that are
already in existence in Europe is that the birds should be
examined visually at least once a day. Have you taken this
into account with your labour requirement? I would suggest
that it is more difficult to examine hens individually in
cages than it is in other systems.

C.M. Hann

I cannot say that I specifically took this into account.
I tried to think in terms of what I take to be current commercial
practice. I agree that staff are required by our codes to look
at birds every day. I would not entirely agree that the cage
system is the most difficult in which to ensure that the birds
are fit and well and that none are suffering. It would be
difficult to carry out as thorough a check in some wire floor
systems as one can make in cages. Admittedly there are features
of the cage system which are not good; the lower tiers of
vertically tiered cages are certainly difficult to see and
probably often get neglected. Similarly, I suspect that multi-
tiered cages present problems. The top tier of a five tier
cage is very difficult to inspect. However, even on range it
is not always easy to inspect the welfare of the stock.

I.J.H. Duncan

I am not suggesting that cages are the worst in this
respect but compared to some systems I think they are at a
disadvantage.

M. Prip *(Denmark)*

I would like to draw attention to the occurrence of
hysteria on wire floors. In Denmark we have had experience of
this; about 7 - 8% of Danish farmers with wire floors have had
hysteria in flocks. That was one of the reasons that the
farmers wanted to change to cages.

C.M. Hann

That is very interesting. If the cage system becomes
discouraged, for example, or becomes less advantageous from a
capital investment point of view, and farmers tend to go for
other systems as a consequence of any action that may be taken,
then this type of problem will obviously figure in the balance.
On the basis of the 'balance of advantage' one could say that
cages have a whole series of disadvantages. However, there is
no perfect system; if we are to encourage the adoption of other
systems we must be very careful that the net benefit is a
genuine one in terms of welfare and not a spurious one that
could be a net loss.

K. Vestergaard *(Denmark)*

On the question of hysteria - Danish experience has
shown that the best production on wire floors is with 10 hens/m^2.
At stocking levels below and at this figure, no cases of
hysteria have been reported.

C.M. Hann

That is interesting; it is the first time I have heard
that the problem was related to stocking density.

K. Vestergaard

It was found that it increased significantly with a
combination of increased density, less space at the feeders
and bad ventilation. It very rarely occurred in litter pens.

C.M. Hann

At 10 birds/m^2 the density of stocking of the house is approximately half that which is achieved under many current cage systems. The temptation must then exist to reduce ventilation and maintain a higher house temperature in cold weather conditions. This would tend to increase ammonia levels and might bring ventilation down to unacceptable levels. One can see that the Danish farmers, wishing to benefit from higher poultry house temperatures, would be tempted to increase the stocking density on the floor to nearer 20 birds/m^2. We know that this level is attainable with proper insulation and adequate minimum ventilation rates to maintain an optimum environmental temperature.

J. Fris Jensen *(Denmark)*

Is it possible to improve the deep litter system in such a way that it can compete with the cage system? The results show that production, or more precisely, the eggs collected, in the floor system is lower than in the cage system. However, comparing the deep litter system with the cage system means comparing 50 year old husbandry methods with modern ones. Nothing has changed in the deep litter system for many years.

C.M. Hann

I am sure it must be possible to make some improvements. In fact, if you look at column 2 in my Table you will see that I have suggested that under good management the eggs collected from a deep litter system can be as many as those collected from a cage system under good management. The asymmetry develops in the litter system under poor management, where eggs are lost or laid in litter, and so on. On the other hand, when you consider egg quality, or dirty eggs, I do not see that the litter system can ever be as satisfactory as the cage system. I realise I am not really answering your question which is a very challenging one. If we put as much effort into evolving an improved litter system as has been put into the development of the cage industry, it would certainly be possible to make progress. But one is up against inherent problems, particularly the undeniable one that in the cage system you are leaving very little to chance, or to the individual variation that exists in the stock. This is really its big advantage from the commercial point of view and the big disadvantage from the behaviour point of view. There will always be problems in free or flock systems due to variability of the stock which do not occur in cage systems. All our endeavours in trying to improve systems other than cage systems will have to be directed at trying to overcome the wide variation that tends to exist in all stocks, however carefully bred.

H.C. Adler *(Denmark)*

There is still a problem with the figures in columns 6, 7 and 8 of the Table. You would have to reduce these figures to 1, or near 1, to compete with the cage system.

C.M. Hann

Yes, because the cage system has emerged primarily as a result of commercial considerations, it is inevitable that a move away from cages will lead to higher production costs, although in some systems they may not be vastly higher.

B.O. Hughes *(UK)*

I know you are very concerned about this problem of getting enough birds in the house to keep the temperature up, and you point to this as one of the principal disadvantages of these systems. However, I also know that ADAS (Agriculture Development and Advisory Service) has considered heat exchange ventilation. Is it possible that this method will become viable before the systems have to be changed on a large scale?

C.M. Hann

It comes back to economics: the value of the feed saved by achieving a higher temperature, against the possible lower investment in other systems which do not allow an optimum

temperature to be achieved. There was an explosion in feed
prices through the 60's, preceding the oil price explosion,
which meant that the industry faced a vast increase in feed
costs with by no means a comparable increase in the returns for
eggs. Fortunately, increased productivity, due largely to the
change to cage systems, enabled the industry to contain the
vast increase in feed price. Circumstances could change. I
know Professor Raymond has ideas about the way this may move
in the years to come. It may be that we will go full circle
and come back to a stage where, in relation to other things,
animal feed prices are less expensive than they are today. In
those circumstances, the premium on maintaining optimum house
temperatures may not be as great; it will alter the whole
equation.

B.O. Hughes

I was just wondering whether we can rely on having a
cheap method of keeping the house temperature up regardless,
within reason, of the number of birds in the house.

C.M. Hann

Well, the problem with heat exchanges so far is that
they haven't worked! But that is an area in which technology
can help us. Another possibility might be to use cheap, low-
grade heat from power stations and so on.

K. Vestergaard

It may also be that maintenance of temperature in the
house is not so significant in some environments as others. As
far as I remember, Danish trials showed that production in a
wire floor house was a little more sensitive to low temperature
than in the deep litter house. If a bird is sitting on wire,
body heat can disperse in all directions whereas if she is
sitting on litter there is insulation there.

C.M. Hann

    This is a very interesting question. One point I should make is that when we talk about temperatures of poultry houses, often we do not define what we really mean. In a battery cage house you can take the temperature at many different points and get a variety of readings. In our work when we talk about an optimum temperature of 21°C, that has been measured just in front of the birds. This implies that with modern cage systems the temperature around the birds is higher than 21°C. In wire floor systems I suspect that the greater sensitivity to lower temperature is due to the fact that there is less crowding of birds and therefore the individual birds are cooler. So it may well be that the optimum temperature, as read by a probe hanging down in a wire floor house, might need to be 23°C or 24°C to give a comparable temperature around each bird. I am only speculating on this, we have not tried it. On litter, it is true you do not have the cooling effect from underneath that you have in cage and wire floor systems, so it could well be that this insulating effect would protect the bird to some extent.

R. Tauson *(Sweden)*

    I wonder if you should include a column in your Table for the keeper, from two points of view, the quality of his environment and the type of work he has to do. I would also like to ask whether you made your calculations with an automatic egg collecting system for the cages?

C.M. Hann

    Dealing with the last question first, no, this was based on hand collection throughout. Incidentally, the saving on labour with automatic collection is not as great as is popularly believed.

    With regard to the environment of the staff, yes, again this is another reason why the industry moved towards the cage system, although it was not anticipated at the time that we

would ever have cage houses containing 35 000 or 40 000 birds -
which is not everybody's idea of an ideal environment. At the
other extreme the environments on free range and semi-intensive
units may be very nice on a spring morning but very discouraging
through much of the winter. Many farmers had great difficulty
in getting staff for these systems, even though they were
prepared to pay extra. As regards the finer details, dustiness
and ammonia levels in deep litter houses are indeed a problem
and have created industrial strife from time to time. Farmworkers are not notoriously strong at banding together and taking
industrial action so usually the problems are overcome at a
local level. But there are problems, and there will be more
problems if more of the poultry industry moves back to litter.

W.F. Raymond

I must thank you all for your participation, the speakers
and those who contributed to the discussion.

MOULTING IN THE DOMESTIC HEN *(Gallus domesticus)* AND ITS USE AND EFFECT

J. Fris Jensen

Moulting or forced moulting in egg production is related to the system of more than one production cycle; some main points in connection with the system in question are described.

PAUSE IN EGG LAYING AFTER THE FIRST EGG LAYING PERIOD

The termination of egg laying is in some instances followed by the formation of a new generation of feathers which pushes the old feathers passively out of the follicles (Lucas and Stettenheim, 1972). This phenomenon is called 'moulting' and is regulated through hormone (Perek et al., 1957), the production of which is influenced by the variation of the length of the day.

EGG LAYING PERIOD

In recent decades the egg laying period of domestic hens has been 12 - 16 months. By regulating the day length in the houses it has been possible to obtain a high egg laying rate, regardless of the seasonal variation of day length.

For economic reasons - the low price of eggs, considerable price differences between pullets and slaughter hens - it may sometimes be necessary to keep a flock of hens for more than one egg laying period (Bell and Swanson, 1974).

R. Moss (ed.), The Laying Hen and its Environment, 259-268

PAUSE IN EGG LAYING - INDUCEMENT AND LENGTH

In order to terminate egg laying for all hens at the same time and to begin the next egg laying period within a reasonably short time, it is necessary to regulate both the termination of egg laying and the length of the pause.

METHODS TO REGULATE PAUSE IN EGG LAYING

Day length, restriction to 5 or 7 hours.

Restriction of water and food (Noles, 1966).

Hormone treatment (7 mg progesterone, injection)(Adams, 1955; Rapp et al., 1971).

Low level of calcium in the diet (< 0.03% Ca)(Gilbert, 1969).

Low level of sodium in the diet (< 0.038% Na)(Whitehead and Shannon, 1974).

High level of zinc in the diet (20 000 ppm Zn in 7 days) (Scott and Creger, 1976).

METHODS CLASSIFIED DUE TO MOULTING

	Pause and moulting	Pause
Day length	+	
Food and water restriction	+	
Hormone	+	+
Low protein	+	+
Low calcium		+
Low sodium		+
High zinc		+

METHODS CLASSIFIED DUE TO HOUSING

Methods	Housing conditions		
	Cage and sloped wire netting	Deep litter	Free range
Day length	+	+	−
Food/water	+	+	+
Hormones	+	+	+
Low protein	+	+	+
Low calcium	+	−	−
Low sodium	+	(+)	(+)
High zinc	+	+	+

INFLUENCE ON BEHAVIOUR

Pause and moulting induced by restriction of food and water.

Crowding by food and water (Lien, 1975).

Higher frequency of agonistic acts per 10 minutes observation (Hembree et al., 1978).

Pause induced by feeding a low sodium diet.

Pecking in feet and toes related to amount of sodium in diet. Frequency related to group size, genetic background and amount of calcium in food (Hughes and Whitehead, 1979).

REFERENCES

Adams, J.L., 1955. Progesterone-induced unseasonal moult in single comb
    White Leghorn pullets. Poult. Sci., 34: 702-707.

Bell, D.D. and Swanson, M.H., 1974. Laying flock performance during three
    production cycles. XV. World Poultry Congress, New Orleans.
    Proceedings: 537-539.

Gilbert, A.B., 1969. The effect of a foreign object in the shell gland on
    egg production of hens on a calcium-deficient diet. Br. Poult.
    Sci., 10: 83-88.

Hembree, D.J., Adams, A.W. and Craig, J.V., 1978. Effects of light
    restriction and amino acid supplementation on performance and
    agonistic behaviour of forced moulted hens. Poult. Sci., 57:
    1143-1144.

Hughes, B.O. and Whitehead, C.C., 1979. Behavioural changes associated
    with the feeding of low-sodium diets to laying hens. Applied Animal
    Ethology, 5: 255-266.

Lien, S., 1975. Reguleret myting i egg produksjonen. Aktuelt fra
    Landbruksdepartementets oplysningstjeneste. Husdyrforsøksmätet
    1975: 67-70.

Lucas, A.M. and Stettenheim, P.R., 1972. Avian Anatomy, Integument.
    Part 1: 197-233. Agriculture Handbook 362, Washington DC.

Noles, R.K., 1966. Subsequent production and egg quality of forced moulted
    hens. Poult. Sci., 45: 50-57.

Perek, M., Eckstein, B. and Sobel, H., 1957. Histological observations
    on the anterior lobe of the pituitary gland in moulting and laying
    hens. Poult. Sci., 36: 954-958.

Rapp, K., Kohler, W. and Schwan, V., 1971. Stresslose Legepause bei
    Zuchthennen durch Applikation von Chlormadinonacetat. 1. Mitteilung.
    Dtsch. tierärztl. Wschr., 78: 542-546.

Scott, J.T. and Creger, C.R., 1976. The use of zinc as an effective
    moulting agent in laying hens. Poult. Sci., 55: 2089.

Whitehead, C.C. and Shannon, D.W.F., 1974. The control of egg production
    using a low-sodium diet. Br. Poult. Sci., 15: 429-434.

DISCUSSION

W.F. Raymond (UK)

May I ask, is the effect of a pause plus a moult the same on total life egg production as just the pause without the moult, or is there an interaction, does the moult itself have any significance?

J. Fris Jensen (Denmark)

I have not seen any figures comparing the systems with regard to economic returns. There might be a difference because by moulting the birds you are giving them a new plumage so the requirements are good for maintenance in the second period. But, of course, this has to be related to the situation of the birds in the two different systems.

W. Bessei (FRG)

With regard to the behaviour of the birds under sodium deficient conditions, we have observed a higher locomotor activity and this would lead to feather pecking. In some trials we have observed an increase in feather pecking which we were able to reduce by lowering the light intensity. We did not observe any feather pecking with the zinc rich diet. With the sodium deficient conditions we had to reduce the day length as well as the light intensity but for the zinc rich method we were able to have a full day length of 16 hours and full light intensity.

W. Goldhorn (CEC)

You showed us some figures on farmers' incomes. It seemed that in every case it was more favourable to moult the birds. If this is the case I wonder why it is not more common for farmers to induce moulting.

Another point, I understand that Ian Duncan has used water and food deprivation to induce frustration. Can we not then say, at least, that food and water deprivation is not acceptable?

I.J.H. Duncan *(UK)*

Food and water deprivation is not exactly the same as frustration. To make the birds frustrated they must have an expectation of food and then be presented with food that perhaps they can see but not obtain. Frustration is a step further than just deprivation. At the same time, food and water deprivation will reduce welfare. I would not like to quantify it but there will be a reduction.

J. Fris Jensen

I also recognise the result of the corticosterone measurement during withdrawal of food and water; the withdrawal of water had the worse effect.

C. Beuving *(The Netherlands)*

Yes, and when we gave the birds water again, the levels came back to normal.

J. Petersen *(FRG)*

I do not think it is necessary to withdraw water and food, if you withdraw food only you can get the same effect. The farmer can also achieve the same effect by allowing the birds a diet of maize meal only.

Can you make a clear division between moulting and the pause? Is it clear, is it significant? When we gave hens only maize meal and water there was a reduction of calcium and amino acids, they had both the pause and the moult.

In your report, you said that another kind of moult is neck moulting. In the laying hen is this connected with depression or egg production or is it simply a feather loss?

J. Fris Jensen

This is rather difficult; there is a rough discrimination between pause without moult and pause and moult. There are variations of moult so it is not clear-cut. It also depends on the method used whether you have a complete change of plumage or a partial change.

I have not found any figures on the relationship between winter pause or neck moult and production but you do find a slight decline in egg production from a flock in which there are a number of birds neck moulting.

On the question, why moulting is not a more common practice, in some cases it is because the egg production is planned years ahead and it is not practical suddenly to decide to have an extra production period in your egg laying flock. You cannot reach the same level of egg production in the second period. It will also vary the size distribution of the eggs, and the egg quality. Unless you have good shell quality in the large eggs you can have very big losses, especially in the highly mechanised systems.

J. Petersen

There is another aspect of shell quality. With increasing egg production there is a tendency to decreasing shell quality in the last months of the laying period. It may be that moulting can be used as a management tool to improve shell quality during that period.

C.M. Hann *(UK)*

There are a number of facets of this which are fascinating. The question has been asked why more moulting seems to be done in the United States than in Europe. One of the reasons is that in California particularly, most of the stock are white egg producing. Certainly there are quite marked differences in the responses of White Leghorn based stock and the heavier brown egg laying strains that we have predominantly in Europe.

But even within some of these strains there seem to be differences in responses to moulting. In England it seems that if the birds moult the subsequent results are better than in situations where they just have a pause. This is particularly true of egg yield, and also of egg shell quality. Normally one expects production to improve after the moult, and then tail away in the same sort of pattern as occurs in the first cycle of production, although the second cycle tends to tail away a little faster.

Turning to pressures in the industry not to moult; there are conflicting interests related to proportions of eggs of different sizes. The larger size eggs from moulted birds can be an advantage or a disadvantage, depending on the price structure for different grades.

It is interesting that moulting has not become more popular over the years and I think this may be related to mortality. It is only in the last ten or twelve years that we have had such spectacular improvements in viability. When this happens it becomes much more likely that a moulting programme will bring economic benefits. Traditional conservatism in Europe is probably another factor. However, there are signs that moulting is becoming more popular, certainly in the UK and, I think, in other parts of Europe.

I.J.H. Duncan

I was in Australia last year and there are very few houses there with a completely controlled environment. They have no control over day length. Moulting is commonly practised there by withdrawing food and water. The usual method is to withdraw food and water for two or three days, and withdraw food for ten days. Of course, they do use a different type of bird; it is a White Leghorn Australorp x; it is a much heavier bird and tends to get fat towards the end of the laying cycle; but nevertheless it seems to me incredible to withdraw food for ten days.

W.F. Raymond

Coming back to the point, is withdrawal of food or water acceptable? Dr. Duncan has quoted an extreme case which one would fully agree is not acceptable. But what about the progesterone treatment you mentioned Dr. Jensen, could this be a viable commercial practice?

J. Fris Jensen

No, I don't think so.

J.P. Signoret (France)

From the point of view of acceptability, when we know the mechanism, leading in a normal situation to a given response in the animal, for instance the effect of a given endocrine balance on a physiological and behavioural response, to my mind it is perfectly acceptable to mimic such a natural event. This seems to me to be preferable to creating stressful situations such as withdrawing food. Controlling the physiological process seems to me to be perfectly acceptable. There is the question of residues but that is a problem for the chemists.

K. Vestergaard (Denmark)

I have two questions concerning behaviour. How common is moulting hysteria?

J. Fris Jensen

I know there have been some comments in Denmark from producers who observed birds which were very active during the moulting period; this was called hysteria. I have not seen any published reports of this.

K. Vestergaard

There has been a Danish study on the effect on behaviour of food withdrawal. It was found that the aggression decreased during the food withdrawal period. It was measured around the feeding trough before, during and after the food withdrawal.

I.J.H. Duncan

It could be just that there was less, or no, competition.

W.F. Raymond

Thank you. We must now move on to the next paper.

CAGES: HOW COULD THEY BE IMPROVED?

R. Tauson

INTRODUCTION

Since 1974, studies of standard cages for layers have
been carried out at the Swedish University of Agricultural
Sciences, Uppsala. The principal subjects being studied are
bird welfare (exterior wear of plumage, foot sores, throat skin
blisters), behaviour (partly by internal video tape recordings),
feed waste and exterior egg quality. Production and feed
consumption are also recorded. About 4 000 SCWLs (18 weeks of
age) of the same origin are put in six different cage
constructions in the same house and given the same feed (15%
crude protein and 11.3 MJ ME/kg feed). The birds are taken out
at 83 weeks of age.

At present, Part V of the study is due to begin. During
Parts I - II, originals of standard cages were bought by the
University and studied. In these initial tests great differences
were found, especially in bird exterior (feathering, foot sores,
throat skin blisters and bird trapping). Therefore, the
manufacturers (the most common Scandinavian, German and English)
were recommended to make some modifications to their standard
constructions which could be tested in a third trial. Our
recommendations were followed to a large extent and, as a result,
improvements were found in Part III. Some of the systems were
taken out and replaced by others. Today, cage constructors
supply new material for further studies to be made at the
University. In some cases ideas for solutions to specific
problems are put forward by the experimental staff.

Many of the results from this project are photographically
documented on colour slides to show the variation in parameters

R. Moss (ed.), The Laying Hen and its Environment, 269-304
Copyright ⓒ 1980 ECSC, EEC, EAEC, Brussels-Luxembourg. All rights reserved

studied, such as foot sores, throat skin blisters and bird trapping. This is also of great help when discussing possible alterations with the manufacturers.

Thus, the aim of the project is to improve cage batteries for layers by comparing details in different types of con- structions and this is done in close co-operation with the manufacturers who regularly visit the research unit. All experimental design and publication is reserved to the University. Reports from the trials are regularly passed out to institutes and other officials and are also very often requested by egg producers.

The research has been focused on the following subjects, which have been found to be influenced by the technical environment:

1) Feather wear
2) Further exterior appearance
3) Trapping of birds
4) Specific behaviours
5) Feed waste
6) Exterior egg quality

In this paper only the first three parameters will be discussed in depth. For a more complete picture of the studies carried out until now, the reports referred to at the end of this paper are recommended. However, I would first like to make a review of the results so far obtained.

1. WHAT DIFFERENCES BETWEEN STANDARD CAGES WERE SEEN IN THE
   FIRST TWO TRIALS?

Feathering

As shown in Figure 1, the feathering of birds kept in cages with solid partitions (cage F) was about 15% better when compared to birds in cages with wire partitions. About 100 birds from each system are scored each 3:d month for feathering

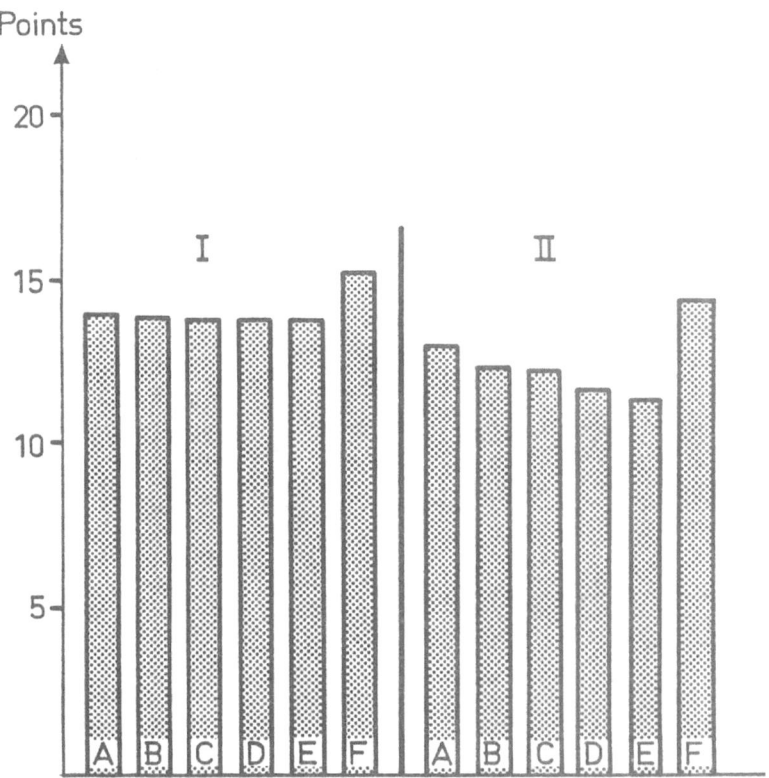

Fig. 1. Average feather score for hens in cages A-F (Tauson, 1978)

on neck, breast, back, wings and tail. Each part of the body
receives points from 1 - 4, where 4 is maximum. Thus the total
points can reach 20.

The differences found mostly concerned a better feathering
on wings and tail and are mainly thought to be due to less wear
against the solid partitions. However, the possible calming
effect on the birds and thus less feather pecking cannot be
overlooked.

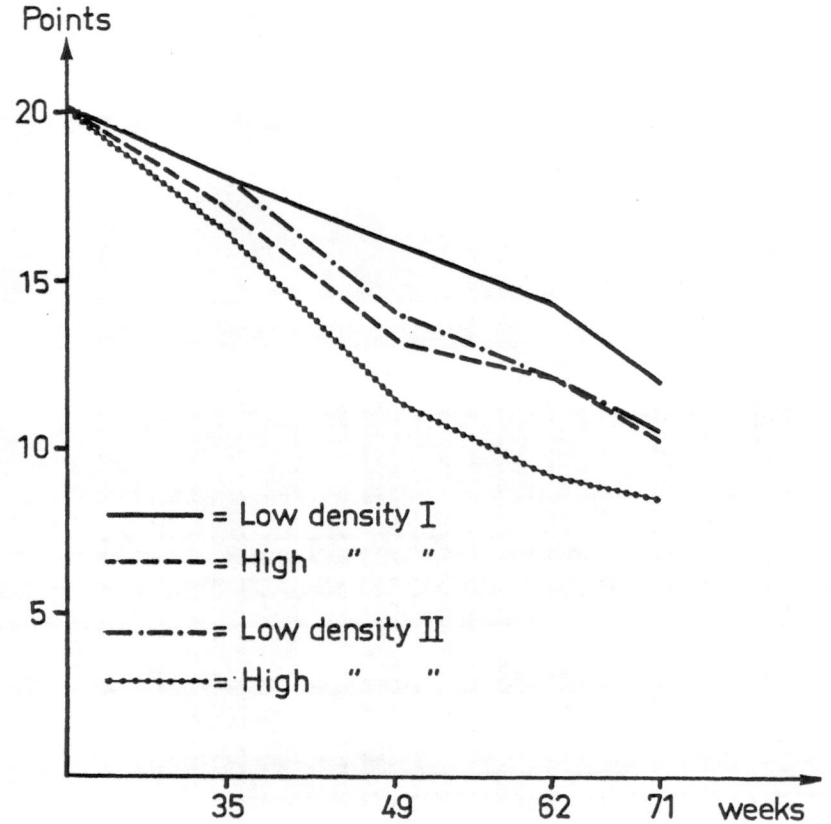

Fig. 2. The influence of age and bird density on feather score
(Tauson, 1978)

An increased density and group size (410 - 460 cm^2/ bird compared to 540 - 620 cm^2/bird, and 3 hens/cage compared to 4, or 4 compared to 5, decreased the feathering by about 15% (Figure 2). It is difficult here to tell whether this was due to density or group size, and whether it was caused by increased feather wear or pecking. The differences were significant until 65 weeks of age and concerned both wings and tail.

Feet health

The variation in feet health (claw fold damage on front toe) between cages was considerable (Figures 3 and 4), as well

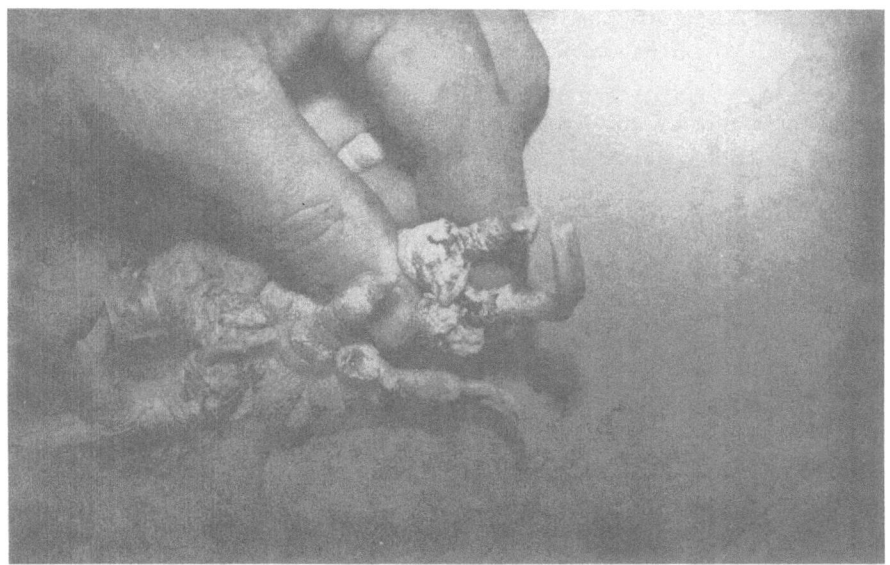

Fig. 3.   A foot of a bird at 65 weeks of age, given 3 points for foot
          damage (see text)

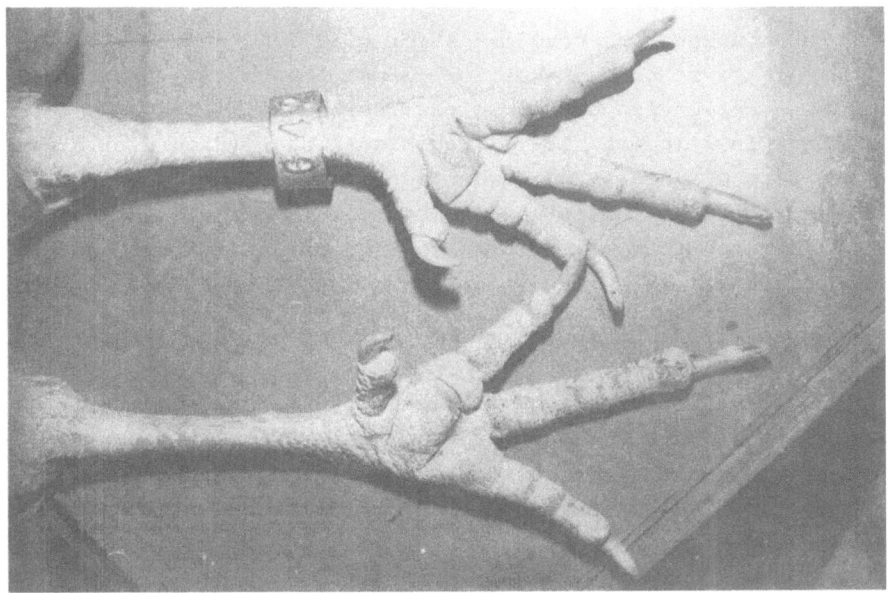

Fig. 4.   A foot of a bird at 65 weeks of age and given 0 point for foot
          damage (see text)

as between birds in the same cage. The scores used every 3:d
month were: 0 points for completely intact matrixes and up to 3
points for severely damaged matrixes. In trial II, 13% of the
birds examined in the worst cage received the score 0 compared
to 79% in the best cage. The main causes of the foot damage,
apart from individual behaviour of birds, were thought to be
the steep slope of the floor and rough galvanizing. At this
stage we also have a beneficial effect on foot health through
the use of epoxy coated cage floors used in one system
(B, Figure 16).

The slope of the floors varied from 23% (13°) down to 14%
(8°) in these initial studies, and the coating from very poor and
rough galvanizing to very smooth epoxy. The floor mesh (1 x 2"
and 1 x 1.5") and wire thickness (2.1 - 3.0 mm) also varied but
no clear correlation with feet health could be determined.
Later (see below) the foot damage on the claw fold was divided
into two main categories.

Length of claws

When using the score 1 - 4 for claw length (1 = extreme
growth of claws and 4 = short or normal length) it was found
that the correlation between clear foot damage (points 2 - 3)
and long claws (points 1 - 2) was significant (Figure 3). One
explanation of this is that when reaching the status of
inflamed foot damage (points 3 especially) there is an increased
blood flow and thus also transportation of nutrients to the
wounded area.

Throat skin blisters

This was a very specific problem in certain constructions.
From Figures 5 and 6 we can see that the variation in status of
throat skin health was considerable. As long as the birds
consume their feed from any kind of trough there will be a
certain degree of wear of the throat feathers. Figure 5 shows
a bird regarded as 'normally' worn at 65 weeks of age. The
stretching length for that bird (from floor to trough bottom)

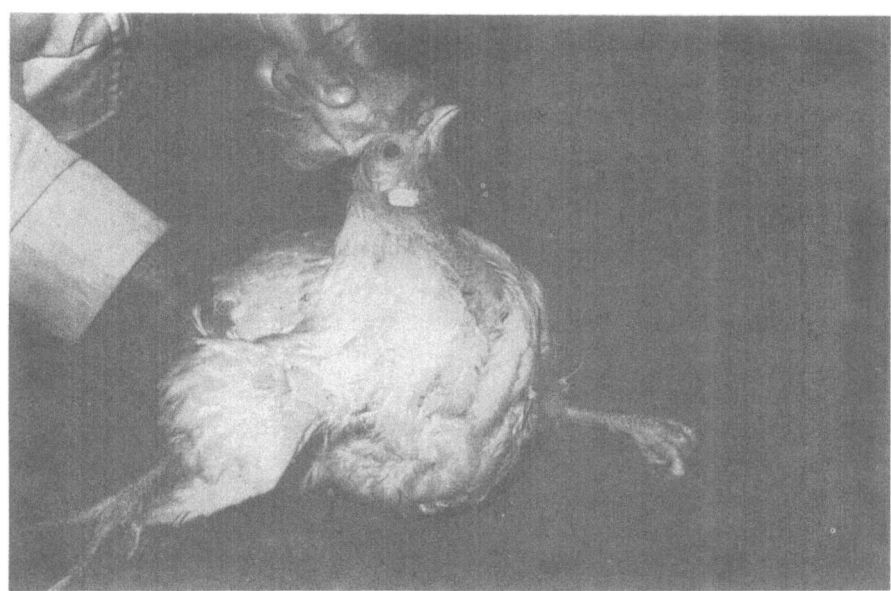

Fig. 5. A 'normally' worn bird at 65 weeks of age (see text)

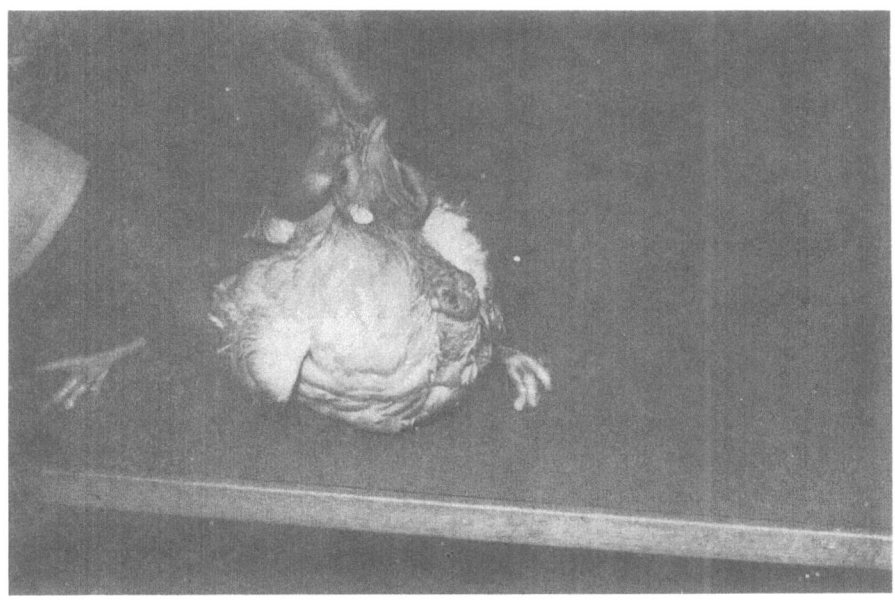

Fig. 6. Throat skin blister at 3 points on a bird of 65 weeks
         of age (see text)

was about 30 cm. The trough also had a very smooth lip.
In this construction no blisters at all were seen. Figure 6
shows a bird of the same age and from the same trial with severe
blisters as a result of feather pins pressed into the throat
skin. In this construction the stretching length was about 30%
greater. The feed trough design was also poor while the lip of
the trough was very sharp. Again, variation between birds in
the same cage (a bad one) was considerable. Therefore the
individual feeding behaviour and the subsequent pressure and
wear on the throat skin are thought to be very important.

It is interesting to note that the feed waste recorded in
the first system described was no higher than in the latter
system. Thus, it is not necessary to make the trough deep or
to install it on a very high level in order to reduce feed waste.
It is more important to keep the feed level low and even.

Bird trapping

Trapping of birds at the neck, comb, wing, leg or toe
appeared surprisingly often in some constructions during Parts
I and II. These cages had complicated locks, uneven spacing
in the fronts, gaps between divisions and/or wire floor.
However, in the more simple and 'closed' cages such incidents
were very rare. Again, the birds are photographed in their
trapped positions and this is of great help when it comes to
discussions of alterations in design of the cages (see
Figures 7 - 10). In Part I (1974 - 1975) an average of 1.7% of
the birds were trapped, ranging from 0.4 to 3.5% in the best
and worst cages respectively. During the next trial this
figure was reduced to 0.8% by alterations in construction.
These alterations were the only ones carried out between trials
I and II. Figures 7 - 10 show some typical cases of bird
trapping in complicated constructions in the initial studies.

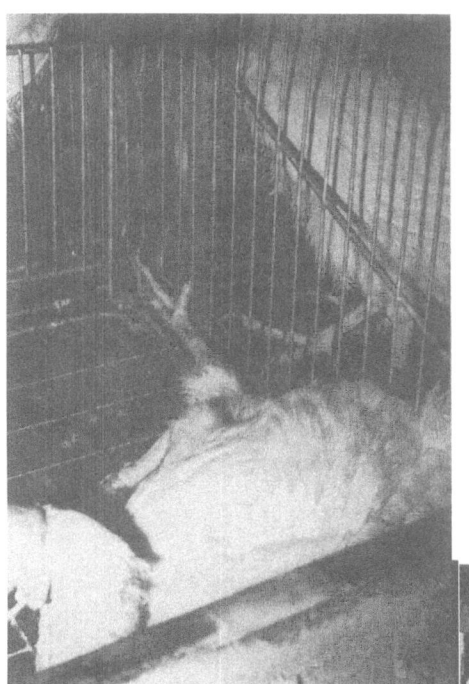

Fig. 7.

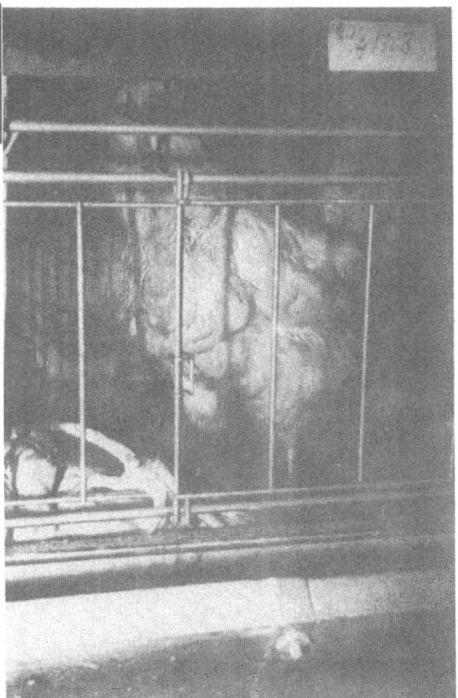

Fig. 8.

Cases of bird trapping in Parts I - II

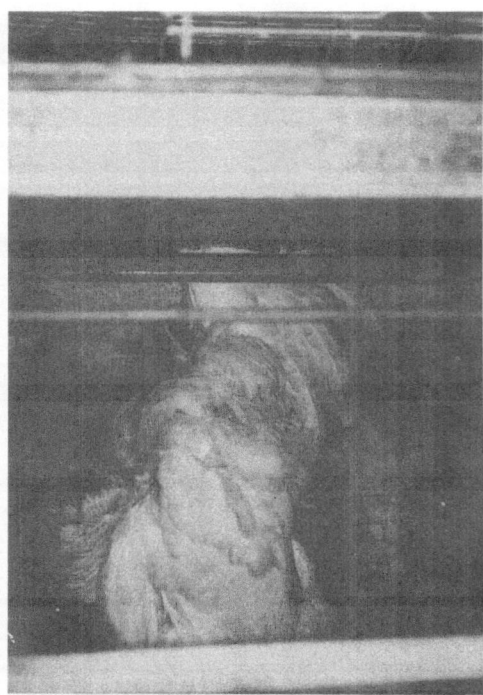

Fig. 9.

Fig. 10.

Cases of bird trapping in Parts I - II

Behaviour

The behaviour studies were focused on feeding behaviour and feed trough space accessibility. By using horizontal bars in the cage front (Figure 11) the birds were able to use the whole feeding space, in contrast to very complicated fronts with locks, hooks or stays where the birds refused to stand (Figure 12). The front shown in Figure 12 was later changed by the manufacturer to the one shown in Figure 13.

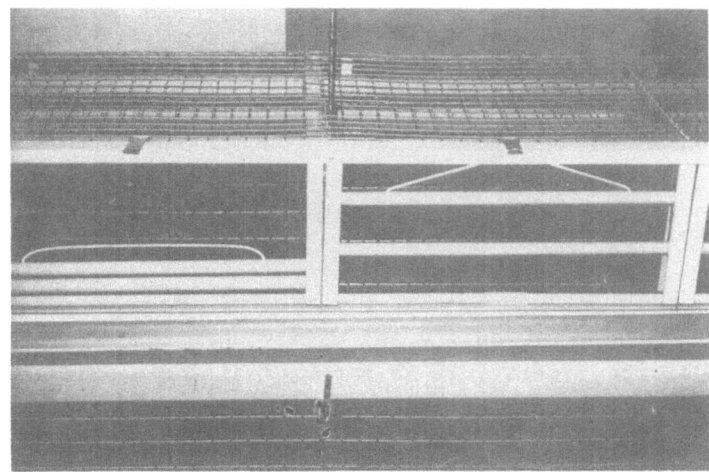

Fig. 11.

Fig. 12.

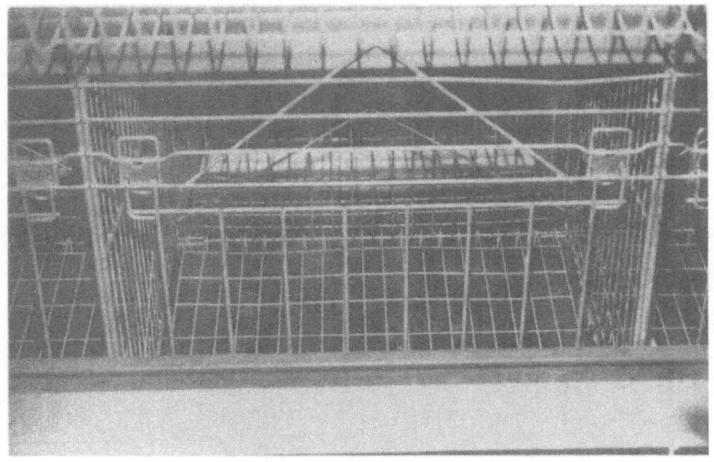

Fig. 13.

Figs. 11 to 13: Front designs providing the birds with different
                degrees of feed trough space

There was also a tendency that birds could approach their
position at the trough more easily and calmly when horizontal
bars were used in the fronts. It is not unusual for a bird at
a vertical front to peck at, or frighten in other ways, the bird
standing in front of her when she is trying to reach the trough.

In the flat deck system where birds feed opposite each
other, pecking on combs across the feed trough was frequent.
The wounds on the combs did not start to heal until the birds
were 60 - 65 weeks of age. It is suggested that this aggressive
behaviour is due to the difficulty for some birds of finding a
position where they are not disturbed by another hen on the
other side of the trough. This feeding behaviour should be
compared to the situation when hens were fed, also from two
sides, at a very long open trough on the floor. In this case
the birds could select a place at the trough more easily because
there was more space. It was also possible for them to establish
the peck order in a mixed flock. It seems that it takes much
longer for birds in a flat deck system to develop a peck order
that includes the birds on the other side of the trough, as
compared with birds kept on the floor. In tier cages where birds

are not in conflict with other birds standing in front of them, pecking wounds were not seen.

General

Technical reliability differed very much between systems, for example, the design of the fronts: how to open them, birds getting in the way, lock design, etc.

Production in g eggs per feed day, was not significantly correlated with the external look of the birds but decreased with the increased group size and density by about 1.5%. This suggests that until a certain limit production cannot be regarded as a reliable measure of bird welfare. Mortality was not significantly affected, except with regard to bird trapping, by group size/density or type of cage.

2. WHAT IMPROVEMENTS WERE ACHIEVED IN TRIAL III?

Feathering

In this trial the construction was changed in two cages to provide solid partitions where previously there had been wire/plastic mesh partitions. Thus, the trial now included three batteries of each kind. The feathering scores of the different cages are shown in Figure 14. The results of trials I and II were confirmed. The new cages with solid partitions improved their feather score and the difference between solid and mesh partitions was again significant up to about 65 weeks of age (Figure 15).

Of course, there are a lot of factors that can affect feathering. For instance, a low relative humidity tends to make the feathers more stiff and consequently they are more easily broken or worn. A low protein level can also decrease the feathering. Light, air speed, temperature, strain of birds, locomotor activity and, of course, the existing hormonal status may also affect feather score.

282

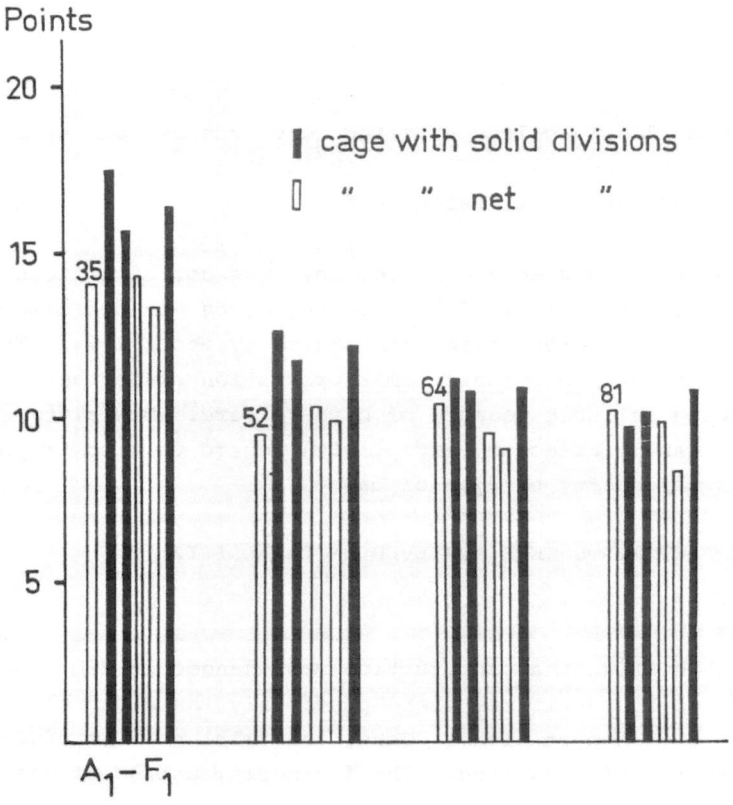

Fig. 14. Feather score (III) on cages A_1 - F_1 at different ages
(35 - 81 weeks) (Tauson, 1978)

Feet health

As shown in Figure 16, the feet health was improved in
Part III. The frequency of birds with completely intact
matrixes (0 points) now varied from 47% to 87% of birds scored.
Birds with 0 - 1 points varied from 70% to 99%. The variation

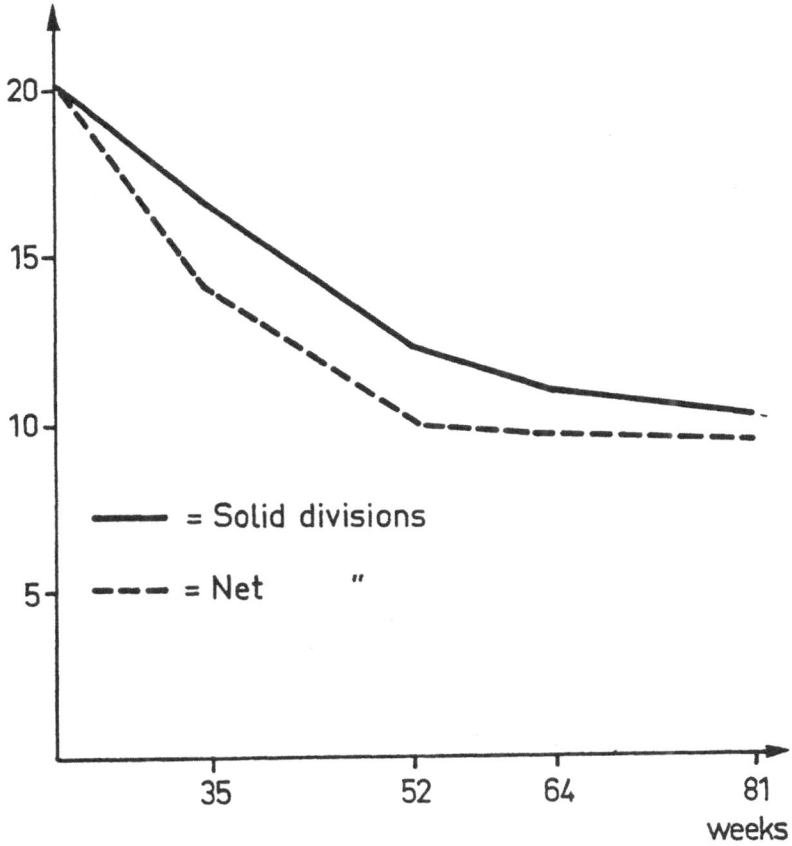

Fig. 15. The effect of cage divisions and age of birds on
feather score (III) (Tauson, 1978)

between certain floors was still significant. The improvement
was achieved in several ways: by reducing the slope of the floors
in systems D and E from 21% and 23% to 18% and 16% respectively,
combined with better and smoother galvanizing and epoxy coating
in system F.

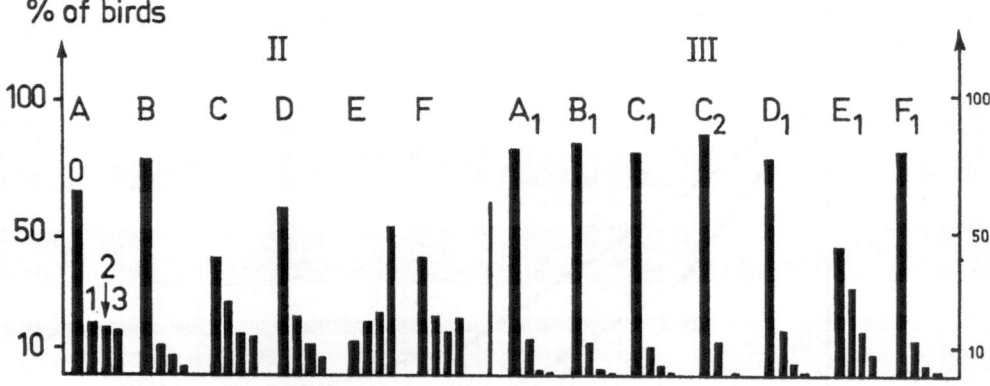

0 = intact matrix
3 = severely injured matrix.

Fig. 16. Classification of scores for injuries on matrix in different
          cages, average of 35 - 81 weeks (Tauson, 1978)

When comparing cage C_1 (epoxy coated) to cage C_2
(galvanized) we can see that there was no significant
improvement through epoxy coating in this cage. This also
indicates that there really are differences between galvanizing
qualities (C_2 compared to F, for example). Possible chemical
irritation to the birds' feet may also be a factor. Since the
quality of the different kinds of epoxy varies in the method
of application to the wire metal, it may be that better feet
health can be achieved on good galvanized floors.

It was found that there are two main categories of
damage to the claw fold (Figure 17, A and B).

A.   Extreme growth of the exterior scales of the matrix close
     to the fold. These defects are seldom inflamed. They
     are thought to be due mainly to pressure and movement of
     the bird's toe in a downward direction. In this case
     both the design of the manure deflector on the feed trough
     and the smoothness and positioning of the exterior floor
     wire are important.

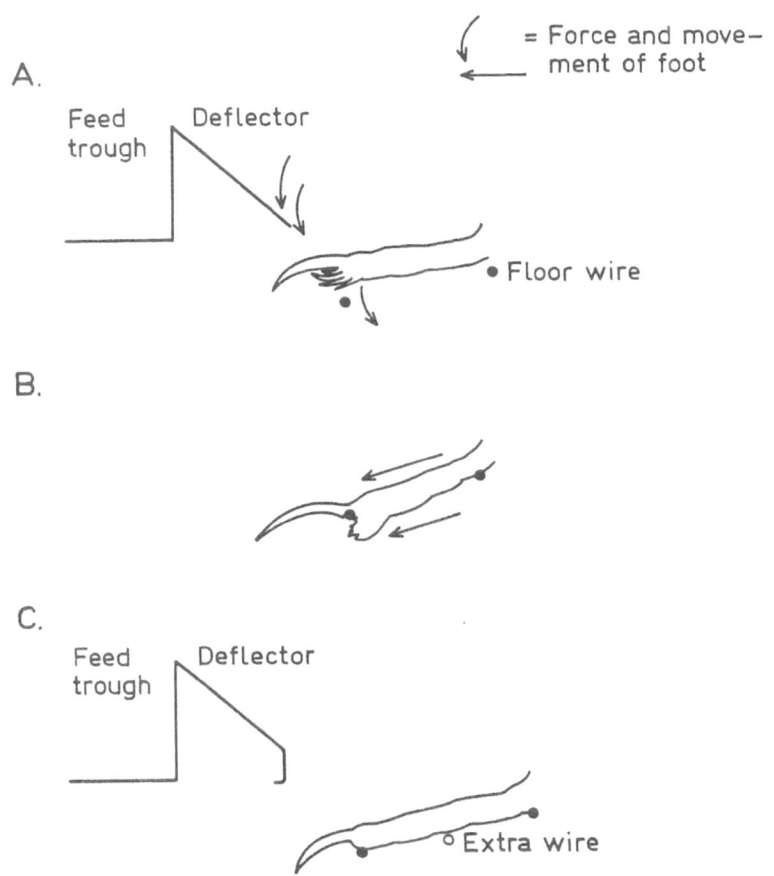

Fig. 17. Theories on feet blisters by laying hens (Tauson, 1979)

B.   <u>An inflamed and swollen claw fold</u>. An impression of the
      floor wire is often seen in the claw. In this case too,
      the surface of the wire is important and perhaps also the
      thickness of the wire. In contrast to damage A, this
      defect is thought to be mainly due to forces in the <u>slope</u>
      <u>direction</u> and thus is affected by the degree of slope of
      the floor; it has little to do with the manure deflector
      design. Much of the improvement in feet health obtained
      so far is due to a clear decrease of type B damage. The
      B type must be considered as the most severe. However,
      combinations of types A and B can also be seen.

As shown in Figure 17 C, an improved support under the bird's foot and a more rounded end on the deflector, might reduce the foot damage even more. Again, it seems that the individual behaviour of the birds is very important. Birds that tend to use these foot movements more than others are far more likely to have wounds than birds which do not move their feet, as described in Figure 17. Therefore, of 4 birds together in the same cage, 3 may have good feet health while one has bad feet health.

Bird trapping

Table 1 shows the results of alterations and new designs during the four trials so far carried out. Starting from trial I with 1.7% of birds trapped, this figure has decreased to 0.17% in trial IV (just finished). In Figures 18 - 25 a review of the old systems of trials I - II are contrasted with the new and better constructions.

TABLE 1

BIRD TRAPPING AS A PERCENTAGE OF HENS HOUSED
In parentheses, number of trapped birds. (Tauson 1974-79)

Trial	Cage						Average
	A	B	C	D	E	F	
I	1.5	0.4	1.5	3.5	1.0	2.5	1.7
	(7)	(2)	(10+2)	(22)	(7)	(16)	
	A_1	B_1	C_1	D_1	E_1	F_1	
II	0.6	0.2	1.1	1.4	0.7	0.9	0.8
	(3)	(1)	(4+5)	(9)	(5)	(6)	
	A_2	B_2	C_2	D_2	E_2	F_2	
III	0.9	0	0	1.3	0.1	0.6	0.5
	(2+3)			(4+4)	(1)	(4)	
	A_3	B_3	C_3	D_3	E_3	F_3	
IV	0	(1)	(1)	0	(1)	(2)	0.17

Fig. 18.

Fig. 19.

Bird trapping - the problem (top) an example of a solution (bottom)

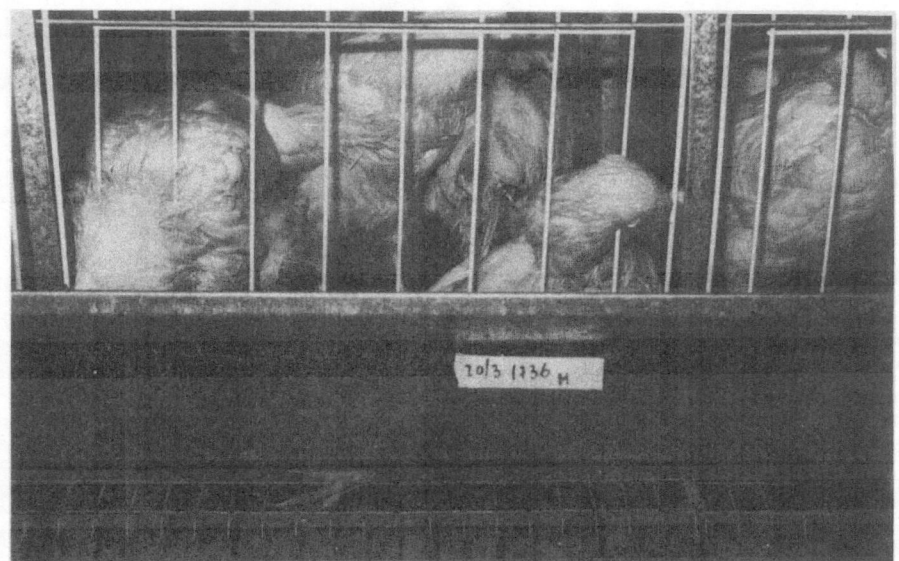

Fig. 20.

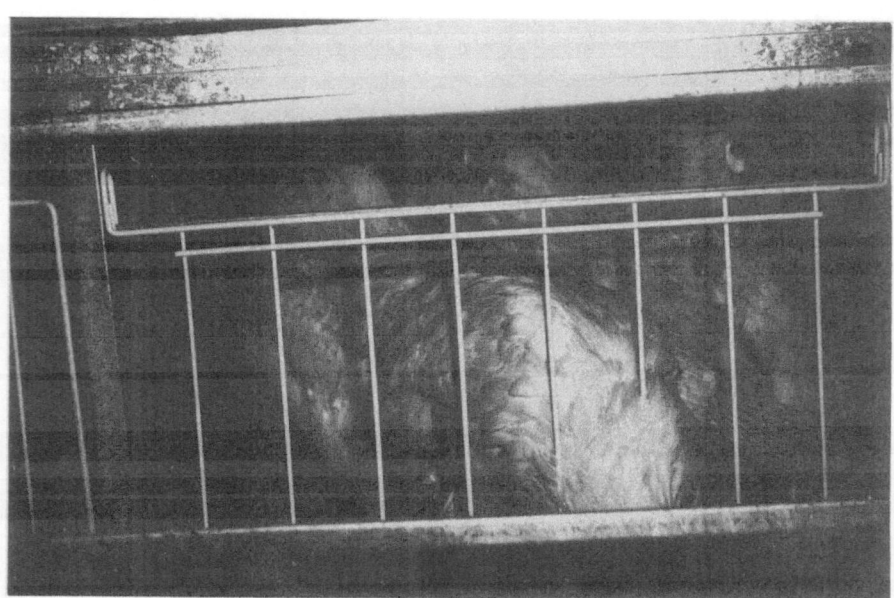

Fig. 21.

Bird trapping - the problem (top), an example of a solution (bottom)

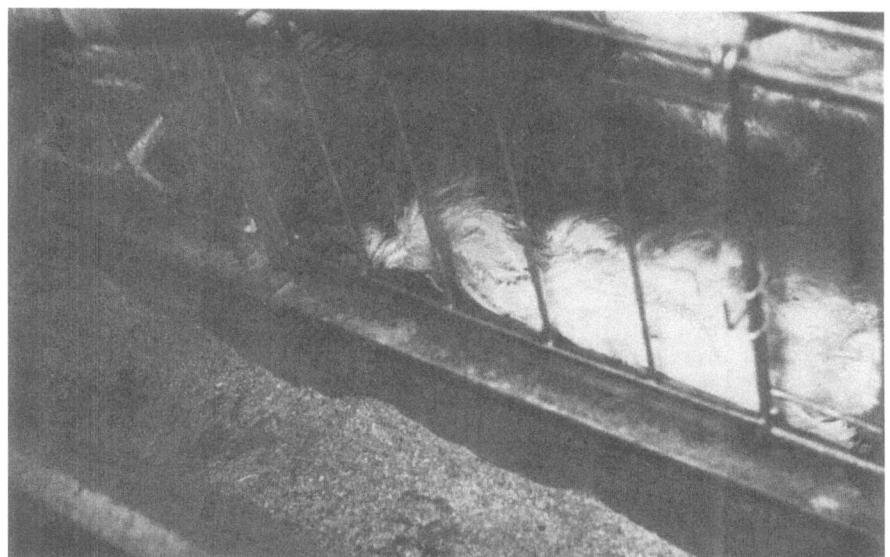

Fig. 22.

Fig. 23.

Bird trapping - the problem (top), an example of a solution (bottom)

Fig. 24.

Fig. 25.
        Bird trapping - the problem (top), an example of a solution (bottom)

When studying the development in this field so far
obtained in the trials, I would like to point out some important
aspects:

1.   The photographs illustrate examples of traps and examples
      of developments and should not be related to any particular
      manufacturer but rather to basic types of constructions.

2.   The degree of improvement achieved in different
      constructions has depended on the magnitude of the
      specific problem in the basic construction. From the
      start some constructions were already much more
      complicated than others (Figure 18 compared to Figure 19).

3.   Trapped birds can be difficult to find unless the birds
      are kept under frequent inspection. Thus, many
      manufacturers participating in the first trials were
      surprised at the results in their own cages.

4.   Bad installation often increases the risk of bird
      trapping, especially when complicated constructions are
      used.

      In our present trial (V) we will keep on working at the
cage design where we think that still more can be achieved.
Soft plastic netting with smaller mesh size will be used over
floors that previously caused feet health problems. This will
also provide more support under the birds' feet and might
reduce the number of cracked eggs. Anti-slip tapes of different
width and material have been placed on some deflectors to try to
shorten the claw length by means of scratching. In an earlier
small scale test at the station this was also tried on perches
but there the tape fell off (Figure 26). However, when put on
the deflectors they stuck to the metal and shortened the birds'
claws very significantly (Figures 27 - 29). More work will
also be done on front designs.

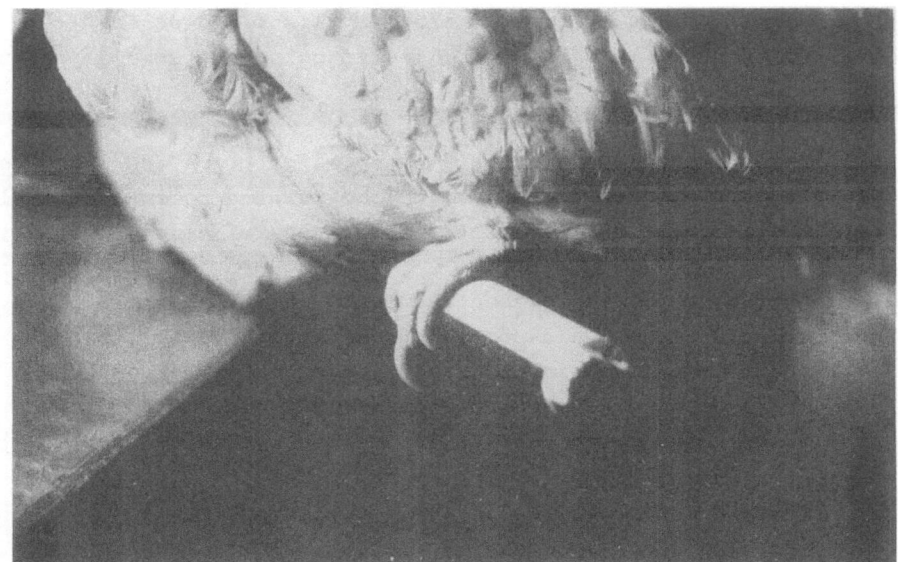

Fig. 26.

Fig. 27.

Anti-slip tapes put on the perches and on the manure deflector
at the trough.

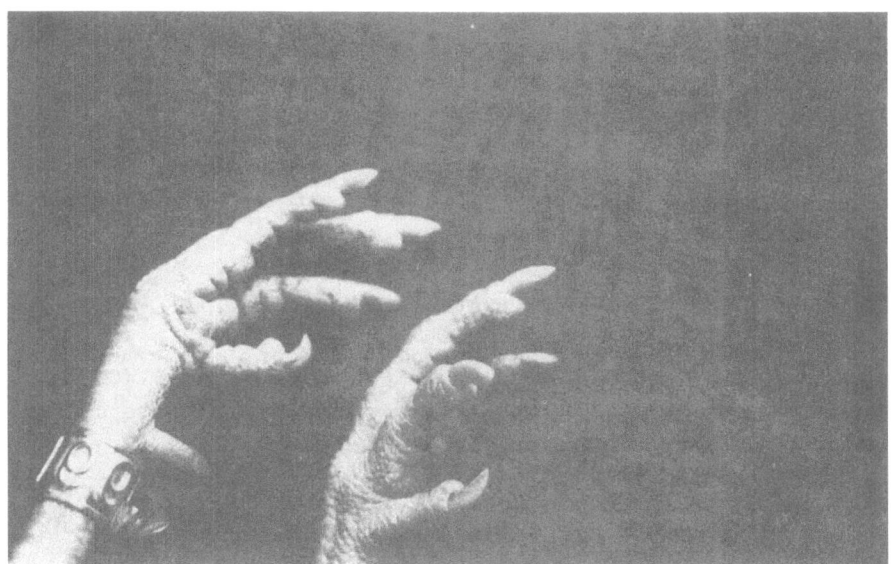

Fig. 28.

Fig. 29.
    Claws of a bird at 65 weeks of age kept in a cage with anti-slip
    tape (top). Claws of a bird at 65 weeks of age kept in a cage
    without anti-slip tape (bottom)

It must also be pointed out that the work on feed waste, exterior egg quality and technical reliability is very important and cannot be completely separated from the welfare problem. Therefore, studies on these parameters will continue. In fact, there are several subjects that must be regarded as of 'mutual interest' (see below) both from the welfare and from the total economic points of view.

3. TRENDS IN CAGE MANUFACTURING DURING RECENT YEARS

Approximately ten cage manufacturers in Europe (Germany, England, Netherlands, Norway, Finland and Sweden) have so far participated in our studies. In looking at cage construction in general during recent years, not only by the manufacturers mentioned above, the principal trends are as follows. In some respects they can also improve bird welfare.

1.   More cages have solid partitions.

2.   A lesser slope of the cage floor is now common. For example, in our first trial the average slope was about 18%, ranging from 14% to 23%. Today, in trial V, the average is about 14%, ranging from 12% to 15%.

3.   Stairstepped cages are more common although the vertical cages are still often seen. This could be important both for better light distribution on the bottom tiers and for easier inspection of birds.

To the above might be added the use of horizontal bars in cage fronts, and more simple constructions. However, this applies mostly to manufacturers already involved in the trials. Furthermore the so-called shallow cages are becoming more common, although mostly in America. Results from studies on these cages are somewhat conflicting. This may be due to differences in the initial feed trough space used in the normal cages in different countries. For example, you would expect a more clear effect in the results when changing over from an initial trough space

of 8 cm/bird to 10 or 12 cm/bird, than changing from 12 to
15 cm/bird.

The developments so far have several causes.

a) More simple constructions of all kinds are often less
    expensive.

b) The deductions for cracked eggs and seconds have risen.

c) The results discussed above have to a certain extent
    already influenced at least the participating
    manufacturers.

4. PERSONAL VIEWS ON SOME SUBJECTS

Although certain very clear improvements have been made
by many manufacturers during recent years, some very primitive
constructions are still found at the big exhibitions. Several
of these have many of the defects described above from our first
two trials in 1974 - 1976, examples being poor galvanizing,
bird traps, etc. My own experience of discussions with the
manufacturers during the last six years is that they differ
somewhat in their way of thinking, although to them the
economics are, of course, very important. However, many of
them tend to have a conservative basic approach to cage
construction; they are often afraid to try something completely
new.

On the other hand the 'welfare lobby' never stops
comparing 'floor keeping' to 'cage keeping' in a very old
fashioned way, and in many countries seldom accepts the
economic and ergonomic arguments.

These two ways of thinking are building up some kind of
barrier between the 'industry' and the 'welfare lobby'. In
this argument I think that both sides forget to some degree that
they also have mutual interests. One very evident example is to
save the birds' feathers. It has been shown by several workers

that the negative effects of bad feathering is very clear with
regard to feed consumption (Figure 30). When the two extremes
are compared - one very badly feathered hen and one very well
feathered hen - the difference in feed consumption is about
30 g in favour of the well feathered bird. This is due to a
better insulation effect. Another example is in cage floor
design, such as more gentle slopes and soft plastic netting.
Both can improve welfare and egg quality.

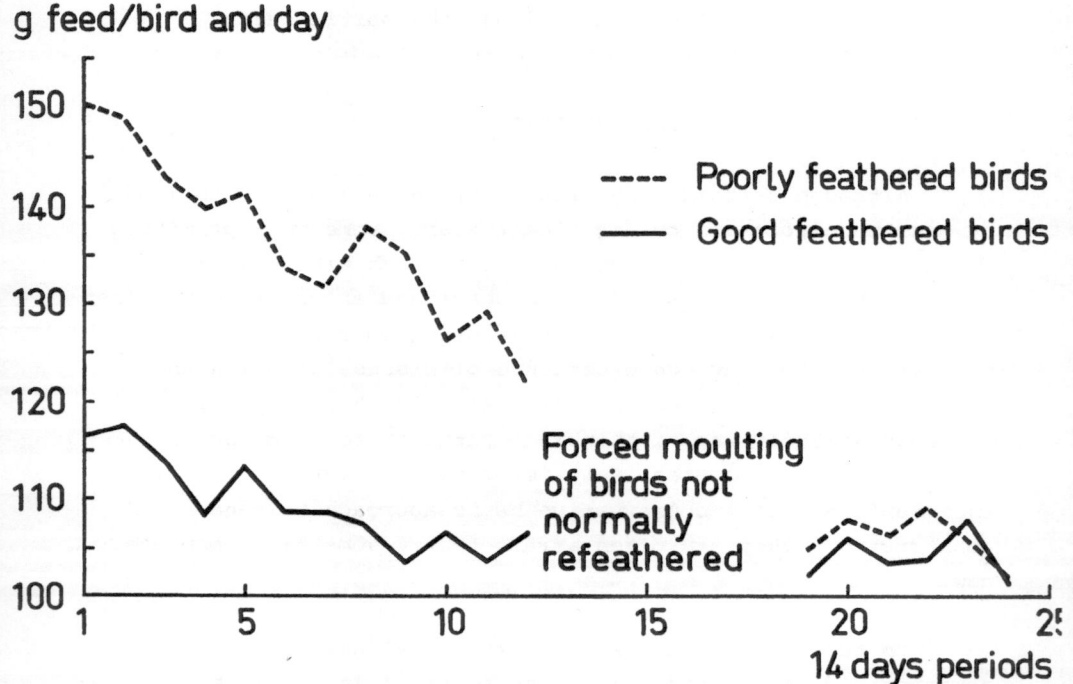

Fig. 30 (Tauson, 1977)

I would like to finish my paper with some personal views
on certain subjects which I consider to be important if we are
to achieve an overall improvement in cages.

1. To get a clear picture of the exterior status of the birds they must be taken out of their cages and inspected closely.

2. Many of today's manufacturers must leave some of their conservative and primitive ways of looking at cage design. They have to accept the welfare problem as such. If the birds are to be kept in cages for their whole lives, the cage design must be more adapted to the birds. In this paper some examples of improvements have been presented.

3. A more close and efficient discussion between the 'industry' and 'welfare research' is important.

4. In countries where the discussion on welfare is blocked between the 'industry' and the 'welfare lobby', both sides probably have to accept a compromise but should also realise that there are mutual interests in bird welfare.

5. There could be a risk in the welfare debate of fixation on two questions:

   A. Floor keeping or cage keeping?
   B. What floor area in cm^2/hen should be stated?

   and in such a debate the work on the specific cage design might 'drown'.

SUMMARY

In this paper I have tried to give a review of the results obtained so far in the studies carried out at the Swedish University of Agricultural Sciences on 'Technical environments for caged layers'. The research has shown that it is quite possible to improve the cage design for the birds. Hitherto the work has been concentrated on an improved exterior of the bird. The importance is pointed out of partly involving different manufacturers as participants in the official research of the University. This means that

improper designs in construction are changed fairly soon.
One example of this is bird trapping.

Clear trends in cage construction during recent years are
discussed and also in what way some of these developments can
improve welfare. Some aspects which, in my opinion, are
important if we are to achieve a more general improvement of
existing cages, are listed and discussed.

REFERENCES

Tauson, R. 1978. Cage design and welfare. Proc. of Overall Assessment of
    Poultry Welfare in Egg Laying Cages. Köge, August 1978.
    Landsudvalget for Fjaerkrae, Vester Farimsgade 1, 1606 Copenhagen V,
    Denmark.

Tauson, R. 1978. Reactions of laying hens to different technical
    environments (I-III). Report No.64 from Swed. Univ. of Agric. Sci.,
    Funbo-Lövsta, 755 90 Uppsala, Sweden.

Tauson, R. 1980. Reactions of laying hens to different technical
    environments (IV). Report from Swed. Univ. of Agric. Sci.,
    Funbo-Lövsta, 755 90 Uppsala, Sweden. In prep.

DISCUSSION

W.F. Raymond *(UK)*

Dr. Tauson, that was a fascinating series of slides illustrating a sequence of improvements. As you say, one really wonders why the people who are in the business of making cages don't realise some of these problems just as a matter of common sense. It is quite extraordinary that it is necessary for research workers to highlight some of them. Of course, some of the things you have been talking about are by no means immediately obvious.

M. Zanforlin *(Italy)*

May I ask whether you have experienced the difficulties which sometimes arise with solid partitions such as problems of light intensity and of young chicks finding the water?

R. Tauson *(Sweden)*

With regard to water, it is just a question of whether the birds have been reared on nipples or open troughs.

As far as light is concerned, there are bulbs today which are constructed in a special way to spread the light evenly over the tiers. Another factor is the colour in the cage. The galvanising makes it very grey and non-reflecting. I am not saying that you want a strong reflection but perhaps white plastic, for instance, might be better.

M. Zanforlin

I would just point out that making an improvement in one part of a system often leads to new problems elsewhere.

R. Tauson

That's quite right.

W. Goldhorn *(CEC)*

You are working largely with plastic material. As I understand it you have had these cages for just one year but it is well known that after several years plastic becomes very rough and hard and may break, especially plastic coated wire. Is there not a danger of injury to the birds' feet?

R. Tauson

As I told you, we had one manufacturer who started to do epoxy floors. The first ones were terrible because the plastic wore off in one month. However, they worked with epoxy for about ten years and we had one cage for four years, throughout the whole experimental period. That epoxy was very good. We have had plastic cages in Sweden for about twelve years. We have not had any problems with the plastic, as such, but there are various qualities of plastic. You know, the plastic people are research workers too, and 'plastic' is a very wide term. I heard there was a proposed regulation in Germany that epoxy coated wire must be used. This is very unfortunate because it would mean that all manufacturers would start epoxy coating and most of them don't know exactly how to do it. I think it would be better to work on improving galvanising, or making complete plastic floors.

We must realise that the floor is not only a problem of feet health; there is also the problem that the manure has to drop through the floor so that the birds won't get it on their feet and the eggs won't get dirty. Also the eggs must roll out, otherwise it is not a practical proposition.

W. Goldhorn

Many governments have regulations, or intend to make regulations, on the degree of the slope. Do you think this is necessary? Will it regulate itself through the necessity of avoiding feet injury and cracked eggs?

R. Tauson

Several factors have been common to cage design in all countries in the least few years. One is that the slope has been decreased. When we started our experiments we had an average slope of 18 - 20%. Some cages had a slope as high as 23 - 24%. Some were as low as 14 - 15%. Today, the average slope would be under 15%. That is the maximum slope allowed in Sweden. So there has been a trend to decrease the slope. Another point is that the vertical three tier cages are not as common today as the stair step cages. One reason for these changes may be the sore feet problem which occurs on wire slope floors but there is also an economic factor, in that the deduction for each cracked egg has risen very much.

There is also a trend for more manufacturers to switch to horizontal bars but here there is a degree of misunderstanding. The main thing is that the birds should have access to the whole area.

J.A. Hill *(UK)*

Have you any comments on improvements to cage gates with respect to getting birds in and out? Some of the existing ones are very badly designed. The second point is improvements to cages with respect to bird inspection - this is obviously quite important from the welfare point of view.

R. Tauson

We have found very great differences in the quality of design in cages. Some cages open inwards, for example, which is very impractical if you are trying to get a bird out.

With regard to inspection, nowadays manufacturers are very much more aware of the welfare aspect and they realise that the keeper must be able to see the birds.

M. Prip *(Denmark)*

I would like to mention one problem which arose during investigations in Denmark. One part of this study was the comparison of slaughter house post mortem inspection results of birds from different management systems. The most striking difference between cage layers and layers from wire floors was the presence of bone fractures in cage layers. Some fractures appeared to have occurred whilst the birds were being removed from the cages or when they were placed in the transport boxes. Others occurred during slaughter, probably by muscular contraction caused by electrical stunning. A large proportion occurred post mortem in the feather plucking machines. There was a very clear variation in the bone fragility between different caged flocks. The interpretation was that there must be some causal factors other than the cage system itself. I would like to ask Dr. Tauson whether he has observed this problem in his investigations and, if he has, whether it can be ascribed to certain types of cages?

R. Tauson

This is in some way connected to Dr. Hill's question about how you get the birds out of the cages. Dr. Prip and I have discussed this before and one thing is clear. The variation and frequency between different farms is very great. We have had between 1 and 1½% of our birds graded as seconds. That means that a bone could be broken, or they could have lesions, or something else. This is a low figure. However, I do know that on farms where perhaps the staff are not very good and the cage fronts are not very good, you see an increase in the number of seconds. That is really all I can say. It might be that birds develop stronger bones in some cages than others - I don't know, but in that case the results are very conflicting. In any case, the birds should be handled as carefully as possible.

L.H. Huisman *(The Netherlands)*

Dr. Tauson, your paper was very interesting but a number of times you said, 'too big', 'too narrow', 'too wide', and so on. Can you translate these things into exact dimensions and requirements?

R. Tauson

No, because there are so many variations. For example, if you have a cage with solid divisions, you can have an extra bird. We should have regulations which would prevent clear overcrowding, if that could be defined. I don't know what the figure is. Our regulations are for 480 cm^2. I would consider that a proper start, or perhaps a solution. Then, you must have a standard cage incorporating this feature and that feature, but it is very difficult. What we do know is the distance between bars, according to whether you use vertical or horizontal bars, the stretching length, and so on. But the area and the group size is definitely very difficult to state. As far as the trough is concerned I would say that you don't need a depth greater than 9 or 10 cm. In the future, when we have more sophisticated feeding systems, it may be possible to go down to 5 or 6 cm. That is one measure. When you talk about the floor it is a complex issue. There is the wire, the material, the floor mesh size, the resilience of the floor, the joints, the stays, and so on. It is very complicated.

W.F. Raymond

I must thank Dr. Tauson very much for his paper.

FINAL DISCUSSION

W.F. Raymond *(UK)*

We are now into our Final Session. We have been
discussing ideas with each other over the last two and a half
days and now we are at the really important stage where we
begin to integrate the ideas we have had and to discuss together
gaps in our knowledge. As you know, there is a Standing
Committee for Agricultural Research in the EEC which is
concerned with co-ordination of agricultural research in our
nine countries. Our Expert Group will be meeting tomorrow to
see if we can begin to define the elements of a co-ordinated
research programme on poultry welfare within the EEC. We want
to draw very heavily on the ideas of this group in beginning
to formulate that programme. So, for the next hour, we want
to have a general discussion on the papers we have had, but
particularly developing your ideas on the needs for research
in the future and the different facets of the subject we have
been looking at.

J.P. Signoret *(France)*

I have noted two points particularly that have come up
during the papers and discussion. The first is that there are
some behavioural problems apart from those leading to physical
injury. When there is physical injury it is simple; it is
objective, measurable, easy to see. The more difficult problem
is in the area of 'frustration' which occurs without any visible
injury to the animal. There are some behavioural patterns which
appear to be eliminated by some husbandry systems, e.g. pre-
laying behaviour, dust bathing. The second point I have noted
is the importance of genetic variability in the way all the
behavioural patterns are presented by the animals under their
system of management. This could lead to three suggestions
for research. First, to stimulate the studies in the field of
the genetics of behaviour, to establish exactly what can and
cannot be changed by genetic selection. Second, we need to
know the criteria which will enable us to measure objectively

R. Moss (ed.), The Laying Hen and its Environment, 305-317.

such things as frustration, the need for dust bathing, and so on. In this connection it might be valuable to make comparisons with other species kept in battery cages, such as guinea fowl, pheasants and quail. Third, we should try to design some objective criteria as an assessment of the quality or efficiency of the design they are proposing. I know in some cases it is not possible but in others it may be so. For example, we could suggest the measurement of pecking frequency in a given situation at a given time of day. This could be used as a reference measure of aggressiveness which could be related to egg production and so on.

W. Sybesma (The Netherlands)

I agree with what Dr. Signoret has said; I think that inter-species comparison is very important. I would also say that it is important to have a comparison between the different approaches such as the ethological and physiological approaches; integration of these might be a new field of combined research.

M. Zanforlin (Italy)

In private discussions with my colleagues during the course of this meeting I have found that the delicate point which keeps coming up is the definition of frustration. It is easy to say that an animal is frustrated in an extreme situation; the problem arises in the range of situations where one might say that mild frustration occurs. Opinions vary between different experts. Obviously it is impossible to provide conditions in which an animal never experiences any form of frustration. If there is a situation in which an animal cannot perform what can be termed its 'normal' behaviour pattern, then can we say whether or not that animal is frustrated? At this point we need a criterion. If the animal performs some other type of behaviour then can that be taken as evidence of frustration? If the animal does not show any sign of any other type of behaviour and does not show any physiological changes, then we cannot say that it is frustrated. Would there be general agreement to this as a definition of frustration?

We might also use a practical measure for the guidance of non-experts - rate of pecking for example, or particular calls that can be easily distinguished.

I think if we start with a definition such as I have described we may be able to arrive at some rough behavioural measure of the frustration, and therefore of the welfare of the animal.

I.J.H. Duncan (UK)

I agree that it is possible to define frustration very precisely. Frustration is a hypothetical intervening variable, but it is possible to define it operationally in the same way that it is possible to define hunger operationally. You can take a lot of different measures and correlate them: the amount of food deprivation, the amount of food taken in in a particular time, the force of pecking, the size of meal, the length of time until the next meal, and so on. You can make a lot of objective measurements. It is a similar situation with frustration. As those of you who are familiar with my work will know, it is possible to frustrate hens experimentally and observe the behaviour that is produced. It is not sufficient just to look for the presence or absence of a behaviour such as, for example, wing flapping in cages, and because wing flapping does not occur in cages, conclude that the bird is frustrated. We now know what the symptoms of frustration are and we can look for those symptoms in cages; if the symptoms do not occur then we can conclude that the bird is not frustrated.

That is the first thing, frustration. However, there may be many other ways in which the welfare of the bird is reduced. I am now working on fear, taking the same approach, seeing what sort of stimuli elicit fear, looking at the responses the bird makes, correlating these behavioural responses with physiological responses.

It is another step to say, well, what about frustration?
How does the bird feel? Is it aversive to the bird? We have
done a little work on this. Severe frustration certainly does
seem to be aversive. If given the choice, the bird will try
to avoid the frustrating situation. By using psychological
laboratory methods it is possible to measure the strength of
the tendency to avoid. So, I think it is possible to get some
idea of what the animal is 'feeling'. Of course, we can never
know what the animal actually is thinking, what its mental
images are, if indeed it has any. But we can see what is its
tendency to avoid particular situations.

G. Martin *(FRG)*

Of course I agree with you that it is not possible in all
cases to say what the animal feels. However, when cage birds
are searching for a nesting place and making dust bathing
movements it is clear that the environment is not right for
them. It is not just that a specific behaviour is prevented
but the birds can be seen to try actively to change the
situation to which they are exposed, by avoiding it and sometimes by trying to get out of the cage. I would say there is
no doubt that the bird misses the nest and the litter. It is
clear that the cage is not the right environment for the bird;
we see it is frustrated and is looking for a possibility to
change this situation.

M. Zanforlin

I think it has been firmly established during this
meeting that birds which are reared in cages show less symptoms
of frustration than those which are put in cages after the
rearing period so, in order to reduce the degree of frustration,
it must be better to rear laying hens in cages from the
beginning.

It may also be possible, as Dr. Signoret has suggested,
to select for strains which do not have a high degree of need
for a nest box, a dust bath, and so on.

There are specific problems which should be included in
the future programme of research.

I.J.H. Duncan

Dust bathing is extremely interesting. It occurs at a
certain level if the hens are in deep litter, or outside in a
dusty environment, and it occurs at a much lower level in
cages, but it does occur. Now, if it actually occurs we cannot
say that the animals are frustrated - they are actually
performing the motor patterns. It might be possible to pursue
this further by using the methods I have described. Could we
not see if the bird will perform an operant response in order
to obtain a dust bath?

Let me take another example, pecking at inedible objects,
say pecking at the cage. It has been argued that this is a
sign of disturbance, that the cage is inadequate in some way.
Well, if we look at a bird in a very natural situation, it will
peck at both edible and inedible objects. Exactly the same
happens in a cage. It will peck at the food; it will peck at
the food without ingesting it; it will peck at the cage. It is
purely speculative to say whether or not pecking at the cage is
a sign of inadequacy.

J.P. Signoret

Pre-laying behaviour appears to be under hormone control.
We know that all the hormone control behaviour in mammals,
sexual or maternal, undergoes relatively wide variations. From
the same endocrine signal they result in very exaggerated motor
patterns such as inter-female mounting in some strains whereas
in other strains it is impossible to see any behavioural sign
of female receptivity. We know that this is under genetic
control. For example, it is very clear that the maternal
behaviour of ewes is completely different in primiparious
and multiparious animals. So there are multiple factors
resulting in very wide variation in such hormone controlled
behaviour.

M. Zanforlin

Would it be possible for us to make a provisional list
of what we consider to be the most important behaviour patterns
that should be looked for, for example, nest searching, dust
bathing, and so on - the types of behaviour which seem to be
most affected in the cage situation, with perhaps, a rough
indication of the degree of frustration experienced in each
case?

R. Moss (UK)

Mr. Chairman, during these last few days I have tried to
evolve a simplified structure of what has been talked about
because I find otherwise that I am getting into such detail
that it is very difficult to sort the wood from the trees.

I have put down on paper a sort of skeleton structure of
what I believe to be involved in each behaviour. I know that
behaviour is a complex thing and multi-factorial but I believe
it can, at least, be divided into learned and inherent; the
inherent behaviour being that which can appear without external
stimulus. Some of that behaviour can be performed, some of it
can only be performed incompletely. The fact that it is not
performed does not really matter because we don't know about the
stimulus. I have put a square around the behaviour which is
inherent, that cannot be performed completely, because possibly
that can lead to frustration or to distorted behaviour. Then
if you go down the other line you find behaviour for which there
is no stimulus. Is that important, that animals do not have
a stimulus for something that they can perform if the stimulus
was present? There is an inability to scratch for food in a
cage because there is no original stimulus. Is that important,
does it lead to frustrative or distorted behaviour? Then, when
you have got the outside releasers or stimuli, there are those
behaviours that cannot be performed, or cannot be performed
completely, and there are those behaviours too where there may
be over-stimulus. Again, does that situation lead to
frustration? Following Professor Zanforlin's point about the

measurement and Dr. Duncan's point about using operant
conditioning to research each of the basic behaviours in the
hen's repertoire of behaviour, is it possible to look at the
behavioural repertoire and, without necessarily grading the
importance of the behaviours, decide that there are one or two
which are very, very important within the present poultry
industry. Nesting behaviour and social space appear to me to
be two that have been discussed during the last few days.

J. Fris Jensen *(Denmark)*

I think we could attempt to set up a model to try to
solve some of the problems. Of course it is possible to set up
a model and to try to describe systems according to behaviour
but I am very conscious of the comment made by Dr. Sybesma
about the different approaches. I am very afraid that the
situation will be described only in terms of the behaviour in
different systems. That is only part of the picture. Then
there is Mr. Moss's intention to describe the situation as
black or white. The few genetic studies which have been done
on behaviour characteristics show it is not black and white.
It is possible to calculate heritability; the level of
heritability gives some indication of the possibility of
changing things. So it is possible to take some, perhaps few,
characteristics of behaviour to put in the sum in order to get
a total picture.

There is also the question of economics - I think my
paper was the only one which dealt with this aspect. We cannot
introduce a system in the poultry industry which would restrict
sales of our products to those in a high income bracket. Our
product is beneficial to health; it is an important addition
to the population's diet because of its protein and fat content,
and so on. Therefore, we must produce it as cheaply as possible.

The point I want to make is that we cannot make decisions
based only on behavioural problems; we have to take into account
the whole picture - I think the people working on behaviour want
to be part of that.

J.M. Faure *(France)*

Heritability has been mentioned as a means to distinguish between the inherent and the environmental. Of course it is a good idea but the problem is that heritability is not fixed for each behaviour. It depends on the strain and on the environment. It is not a foolproof measure of what is due to the environment and what is due to the genome.

W. Bessei *(FRG)*

That is true but I think heritability is a good indication of what we can do genetically to change behaviour in a required way. If we have genetic strains which show very high frustration in behaviour patterns, pacing for example, and the heritability of this special behaviour and the genetic variability are very high, then it would be possible to reduce the level of pacing by genetic selection.

C.M. Hann *(UK)*

I would like just to come back to a concept which was mentioned yesterday by Dr. Duncan, that of social space. At the moment this is a vague concept and I wonder whether some kind of research into this could discover thresholds below and above which behaviours differ. This would not necessarily indicate that one behaviour pattern was better or worse but if it showed differences then this would tend to support the idea of a social space and might give guidance to spaces that might ultimately be felt to be desirable in a cage, or indeed in other situations.

Another area which may be interesting to look at is the range of stimuli that the bird experiences in all situations but particularly in the cage situation where in some directions the experience of the bird is restricted by its environment. An idea which has been mentioned and which may be worth pursuing experimentally, is that of enriching or changing the visual, audio or tactile aspects of the environment to provide interest for the bird and, as a result, alter its response to potential

frustrations. It has been the practice in some poultry units
to have music. I don't know whether anyone has looked at this
seriously from a behavioural point of view; it has been looked
at from a commercial point of view and it has been claimed to
have helped.

J.A. Hill *(UK)*

We have concentrated our discussions on the laying hen,
the adult stage. I think we might profitably explore the interactions between the rearing stage and the laying stage with
respect to behaviour bearing in mind that they are very often
quite separate. Certainly in the UK there will be a rearing
unit and then the birds will be moved to a completely different
place for the laying period. With all animals, a lot of the
learning occurs in the very early stages and this may well have
a considerable effect on what will, and will not, be acceptable
to the bird in the later stages. So, I think we should look at
the thing as a whole, right from day-old; the interaction is
quite important and is often ignored.

H.C. Adler *(Denmark)*

I would agree with most of what has been said. However,
there is one thing which we must realise: research on these lines
will not benefit the commercial hen for many years, and she will
still be in a cage. Dr. Tauson demonstrated this morning that
much can be done to reduce the occurrence of many of the causes
of injury etc. Is there any possibility that something could
be done at EEC level to ensure that better cages are constructed?
Could there by an initiative from Brussels on this point?

K. Vestergaard *(Denmark)*

I would like to make two comments. Firstly, with regard
to genetic selection, I think it will be very difficult to
select for a number of desirable traits and against a number of
undesirable ones. The first priority has to be to select for
good egg production and this may well suffer if the geneticists

try to select for and against various other traits. I would
suggest we concentrate on the environmental problems that we
know exist; the hen is there now, selection is in the far
future.

My second point is about social space. I don't think we
have given sufficient consideration to systems other than cages.
For example, in Denmark, provided the producers obey the law
and allow 600 cm^2/hen with 3, 4, 5 or 6 hens per cage, then the
difference in production costs of an egg is only 1 - 2 øre,
i.e. 2 - 3%. That is within the normal range of variability.
So, if this meeting concludes that we cannot accept a stocking
density greater than 600 or 650 cm^2/bird, then it will stimulate
development of other systems which have many advantages and a
greater potential for improvement. For example, if you have a
wire floor, it may be possible to provide a small dust bath;
the hen only requires the dust bath for a limited time each day.

W. Goldhorn *(CEC)*

I would like to reply to Dr. Adler's question. It is
extremely difficult for the Commission to put forward proposals
which would push the industry to accept Dr. Tauson's findings,
for example. He is not even sure himself what would be the
best design for the cage; he needs more time to research it.
It is the same problem with stocking density. We do not have
a definite area below which we can say suffering begins. It is
extremely difficult to do anything until there are definite
figures.

H.C. Adler

It was not a question of measurements in centimetres; it
was a question of the quality of the cage, correct construction,
the avoidance of obviously poor construction. Can you not
impress governments to take steps to ensure that within their
own countries a similar cage is made as in Sweden by co-
operation between a university unit and the industry?

L.H. Huisman *(The Netherlands)*

In addition to the points already raised I think more research is needed into improving existing systems of housing and keeping animals and into developing new systems of housing and keeping animals. I believe these two items should have more attention than they have up until now.

R. Tauson *(Sweden)*

I would like to explain briefly how the system works in Sweden. When a farmer wants to construct a unit for hens he must have the proper area for the manure and so on. He has not only to show plans for the house but also for the cages. Veterinarians from the counties frequently call us to ask advice on particular types of cages which indicates a sparse knowledge about cages among the health authorities. However, they do keep in contact with us. If the word gets round that a certain type of cage is liable to damage the birds then the farmers don't buy it. There are encouraged to look at alternative cages, and even alternative systems. Many of the egg producers come to us for advice. The effect has been that several cage manufacturers who sold cages six or seven years ago cannot sell their cages today. But that has not been done by force - it has been done by scientific work.

W. Bessei

With regard to behaviour traits, I would support Mr. Moss's proposal to make a check list of the different traits and to see what information we have about the welfare of the animal in relation to each one. We have tried to do this in Germany with a group of ethologists. I have to admit that with most of the traits we could not agree; we could not say whether increased wing flapping is a sign of more or less welfare. However, there are some behaviour traits where 99% of the ethologists are convinced that there is a definite effect on welfare. We should check to see in which behaviour traits the situation is clear and in which more work is needed.

W. Goldhorn

Would it be possible for you to publish a list each year, as the consumer groups do, or the automobile clubs, comparing the different types of cages according to security and so on? This type of pressure would be preferable to government intervention.

R. Tauson

In Sweden the names of the manufacturers are mentioned in the reports but we regard them as systems. Something along the lines of what Dr. Goldhorn is suggesting has been discussed in our country with a view to having some kind of certification of cages. However, we are afraid that such a system would arrest further development; there would be a danger that it would be accepted that all cages should be like this in the future. As it is, we never know - perhaps in two or three years we will have a cage which is made completely of plastic, or rubber. We must not fix standards while there is still room for improvement. Perhaps we should have certain recommendations or regulations regarding space - that might be very important.

J.P. Signoret

I do not entirely agree with your argument about fixing standards. For example, in France there is an official agreement for new seeds for cereals, grasses and so on. When a new strain is developed it has to be proved to reach certain standards before it is allowed on the market. This certainly does not result in fixing. The seed companies are still developing a large number of new varieties. The advantage to the farmers is that they know exactly what the performance of a new seed will be. I think that such a system for cages might be a very good thing.

D.E. Hood (Ireland)

As an observer rather than someone who is familiar with the field, I wonder if it would be possible to establish some index of welfare which would attempt to quantify systems but leave the individual elements without definition.

J. Fris Jensen

On the subject of laying down standards for cages, if we look at the poultry industry and take an example of the random sample tests which have been run for 50 years now, they have had an enormous influence on the strains, or strain crosses available to the market. The testing of systems is only four or five years old but in my opinion it has already had an enormous influence. So I think it would be wise to continue such work in order to get rid of all the bad systems - the sooner the better.

I.J.H. Duncan

May I ask Dr. Hood if he can give an example of the sort of thing he means?

D.E. Hood

Well, as I see it a lot of the elements can be measured. As you said yourself, objective measurements can be made of many of the traits. What I am suggesting is that these should all be combined to form an overall index of welfare.

W.F. Raymond

I should say that the answer is no!

I.J.H. Duncan

Yes. Theoretically it might be possible but practically there are so many things which influence welfare that I have a feeling it might be meaningless.

SUMMARY

W.F. Raymond *(UK)*

I have been asked to summarise the discussions both of
this session and of the conference; I am perhaps an odd choice
for this task, for I have never worked with poultry and I am not
a veterinarian, but a ruminant nutritionist. Yet I was probably
working on aspects of animal behaviour earlier than quite a few
of you, studying the behaviour of cattle and sheep grazing in
the field - in fact my team was probably the first to fit radio
transmitters on beef animals which signalled back to base a
record of what the animals were doing. So, although I know
relatively little about poultry, I do have some longer term
experience on animal behaviour, and in particular on the problems
of statistical analysis and interpretation of this work. This
has arisen many times in our discussions, of getting adequate
replication in behavioural work because of the real difficulty
in taking enough measurements - as in the work reported on
hormone status. I suspect that some of the results which have
been reported here, and which we have been discussing, may not
reflect as clear-cut differences as we may have assumed.

A second important problem which has emerged from our
discussions is that of adequate replication in time. In other
words, a well-replicated experiment may give very clear results
at a particular time, in a particular laboratory and with a
particular strain of bird, but another experimenter may get
quite different results elsewhere. In fact this should not
discourage us, because the critical study of results from
different workers may frequently be just as valuable as the
detailed analysis of results within a single experiment; it is
often the discrepancies which emerge between well-conducted
experiments which lead to new concepts which would never have
been obtained from the single experiment.

R. Moss (ed.), The Laying Hen and its Environment, 319-326

320

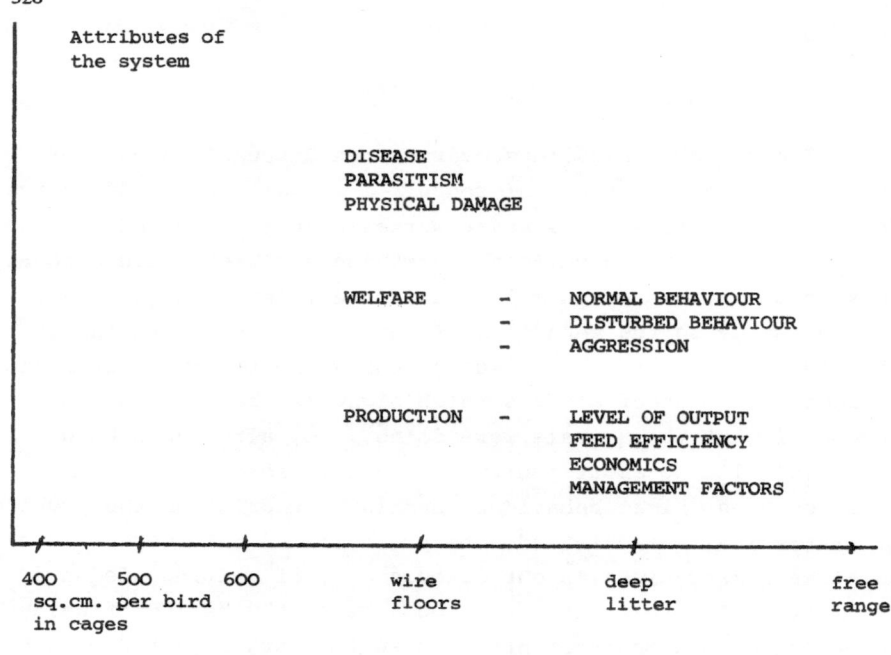

Fig. 1. The interactions between systems of housing egg-laying birds and
         attributes of the systems.

After the meeting yesterday Dr. Martin suggested that we
might summarise some of our discussions graphically, and I have
done this, in Figure 1, which examines the likely consequences
of a steady decrease in the intensity of egg-production systems
(on the horizontal axis), and a series of criteria of 'success',
in production, behaviour, and other terms, on the vertical axis.
An interesting point that then emerges is the little attention
that has been given to production efficiency in our discussions.
But the question was raised, particularly by Dr. Signoret - and
we have not answered it - is high production an index of adequate
welfare? My own impression, working with other classes of
livestock, is that it might be risky to make either conclusion -
that high productivity necessarily reflects adequate welfare
conditions or that low production indicates that we are ill-
treating animals. Thus consider store-feeding of beef cattle -

cattle moved from summer grazing into housing, where they get
a low level of nutrition during the winter so that they just
about maintain their weight - and are then in fine condition
to make very high rates of gain from grazing during the following
summer. I would question whether there is any evidence that
such cattle are under poor welfare conditions during maintenance
winter feeding.

Equally the same cattle could be fed on a concentrate
diet to achieve and make very high rates of gain; but as our
experience with barley-beef feeding in Britain shows, there
can be considerable risk of liver damage with intensively-fed
cattle; I would not suggest that liver damage reflects either
good or bad welfare, but it does certainly indicate a deviation
from normal in these highly-productive animals.

Another point which has been raised a number of times is
the question of disease and parasitism in poultry managed under
different systems. I strongly believe that an animal which is
diseased or parasitised indicates a probability of poor welfare
and that we should be at least as concerned about birds being
diseased or parasitised or physically damaged, as that they may
be frustrated in certain systems. Thus we need to look very
carefully at the spectrum of possible systems in Figure 1 to
consider how effectively each allows us to control disease,
parasitism and physical damage, and to question whether there
is adequate welfare in a system in which the birds may show all
the normal behavioural patterns but are at the same time
diseased or parasitised, and so suffer discomfort. Thus it was
particularly interesting to learn from Dr. Tauson's paper that
there are sensible and practical measures which can be taken
to minimise physical damage in battery cages - and surprising
to learn that, despite this information, cage manufacturers are
still producing cages that allow birds to be damaged in the ways
he illustated. Surely this should not need legislation; the
word should quickly get round the farming community that a
particular make of cage damages birds and should not be used.
Thus I would agree very much with Dr. Jensen that what is needed

is the spread of information, through the media and the
extension services (rather than by government or EEC edict)
on the structural characteristics of cages, or of deep litter
systems etc. which will minimise physical damage to birds, so
that we create a climate of opinion in which manufacturers are
only able to sell equipment which satisfies these requirements.

There is much that can be done in all types of egg-
production systems to reduce risks from disease, parasitism
and physical damage; but certainly none of the evidence
presented to us would convince me that the more extensive
systems which are put forward as alternatives to the battery
system are yet satisfactory in these respects. In particular
I would question the concept that the most 'natural' environment
is necessarily the optimum environment for animals. For in
domesticating livestock we have consciously decided to eliminate
as far as possible those features of the natural environment
which are 'red in tooth and claw'. In the process we may have
also limited the animals' ability to exhibit what we (sometimes
subjectively) consider as desirable behaviour patterns; but on
balance I believe most animals suffer much less than their
cousins do 'in the wild'.

As Dr. Hood has suggested, it should be possible to list
and roughly quantify the importance of a range of 'desirable'
features such as social space, nesting behaviour, dust bathing
etc. But the first priority that has emerged from our
discussions is the need to prevent disturbed behaviour, and for
me this is possibly more important than insisting on a complete
portfolio of normal behaviour. In other words although full
normal behaviour may not occur, as long as it is not replaced
by a disturbed behavioural pattern there may not be great cause
for concern.

Again I have been very interested by the indications in
our discussions that it may be possible to modify behavioural
patterns, hopefully in a desirable direction. In this we
concentrated in particular on modifying behaviour by increasing

the size of the cage, by changing the number of birds in the
cage, and so on. But Dr. Faure raised the possibility of
genetic change; should we modify the genotype of the bird to
adapt it to an existing environment, and if that environment
is considered by some as inherently unsatisfactory, is it then
ethically right to modify animals to adapt to it? This does
of course perhaps beg the question, because it assumes that
there is an agreed definition of an unsatisfactory environment.
I think the concensus of our view was that, while it is
important to examine, within the existing genetic diversity,
whether certain strains of birds are better adapted to certain
environments than are other strains, that we would not give
any priority for breeding towards this end - not from any ethical
considerations, but because there are likely to be more
promising methods of improving the welfare status of poultry.
Further, one might well be selecting birds for an environment
which in future becomes unacceptable in both welfare and
economic terms. So, I would give main priority to the study
of the methods of modifying behaviour by changing space and the
physical design of cages etc. to ensure that birds can live and
feed without physical damage and disease. Most interesting is
the concept which Dr. Hill mentioned, that birds reared in a
particular environment may adapt better to a similar environment for their productive life - that birds reared in cages
will adapt better to production in cages than birds reared on
free range. I wonder what would be the interactions between
two birds reared in cages and two birds reared on deep litter
all housed in the same cage? This leads on to another point
made by Mr. Hann. If we do not insist on the full spectrum of
normal behaviour (and how can we, as we do not know what it is?);
and if we put more emphasis on avoiding disturbed behaviour,
can we improve the birds' environment by making it more
interesting? This could be by providing nest boxes, dust baths,
etc. in an attempt to stimulate a 'normal' behavioural pattern
but as we have been told this can result in further problems
such as egg-laying in the dust bath. But perhaps we need to be
more imaginative, by creating interest in ways other than just
trying to mimic the 'normal' environment - for instance as in

recent studies on introducing coloured objects into the cages.
I would be interested in the concept that each bird in a cage
might have an object to which it was adapted in the rearing
stage, which is unique to it, and which differs from the
interest objects of the other birds in the same cage. Thus,
if a cage group was made up of birds which had been separately
reared with green, red, blue or white objects, each bird in
that cage could have its particular coloured object as its own
interest object; might this reduce aggression within the group?

Finally, while we accepted that it would not be a subject
for discussion at this Conference, at the end of the day we must
be concerned with economics, and the need in practice for some
compromise between production, welfare, and the profitability
of different systems. As Dr. Brantas said, there is no such
thing as the completely satisfied animal; in the same way there
is not a completely healthy (diaease-free) animal, and we have
to consider the interactions between production costs,
satisfaction, health status etc., in different systems. Our
responsibility as research scientists is to obtain as much
reliable information as possible to allow extension workers,
manufacturers and farmers to develop and install systems which
allow an acceptable compromise between profitability (and so
the price of eggs to the consumer) and the often ill-defined
concepts of animal welfare that we have been discussing.

I would question whether we have been tackling this
problem in the most effective way. At this meeting we have
representatives from something like twenty research teams
working in our different countries, in too much of the work
that has been reported I believe the size of these research
teams has been too small to make an effective contribution to
solving the problems we have posed. As Dr. Sybesma noted, in
an experiment we should not just measure behaviour, but other
parameters as well and too often staff are not available to do
this. Dr. Wegner said that the scope of her work had been
limited by shortage of the staff needed to take the data
essential for proper interpretation of her observations - and

that there is a danger that she may not have any staff at all
later in 1980. I wonder whether, within the co-ordinated
programme of research in the EEC which we shall be discussing
this afternoon, we should not be putting much more emphasis
on making more effective use of the considerable overall
research capacity and capability that we have between us, but
which, because it is spread so thinly, means that few of our
centres are carrying out comprehensive experiments from which
valid conclusions can be drawn. This is in no way intended as
a criticism of the smaller teams, but to question whether, without a full range of expertise, or adequate numbers of assistants,
it is possible to do valid work in this difficult subject of
animal welfare and behavioural science. In the EEC discussions
we must first aim to decide what are the real priority subjects,
and having done this, to see whether we may be able to tackle
these subjects in a co-ordinated programme rather than
individually - and hopefully including workers from Sweden
and Switzerland, even though they are not in the EEC. A key
to this could be in being able to move specialist staff, more
readily than is now possible, to work in each other's laboratories. At present each individual laboratory applies its
own limited range of disciplines and techniques to its own
limited experimental facilities, so that scale and replication
are inadequate, an incomplete range of data are collected, and
interpretation of the results is compromised. I believe we
have a choice; for each laboratory to continue as at present,
carrying out fascinating experiments but with insufficient
replication or range of expertise; or for us to set up a
smaller number of larger experiments, with specialists in
different disciplines from other laboratories seconded to work
in a team large enough to collect the comprehensive data needed
to understand this complex subject of welfare and behaviour.

A new aspect of 'scale' of work was raised by Dr. Hill,
who noted that behaviour is likely to be affected by scale of
operation: once we move from studying battery cages (where the
cage can be both the experimental and the commercial unit),
Dr. Hill suggested that small-scale experiments on systems such

as wire-floor, deep litter or free-range may well produce
results that would not be valid on a commercial scale. That
is, the experimental unit may be so much smaller than the
production unit that it is not possible to extrapolate from
experiment to practice. If this is so then it could mean
that experiments on these alternative systems must be carried
out with large groups of birds if the results are to have any
meaning in behaviour and welfare terms - and few individual
research centres are able to support research on this scale.

Unless this problem is tackled it could mean that we are
obtaining relevant results from studies on cage systems, but
are drawing faulty conclusions from studies on wire floor,
deep litter and free range systems. If we are to make
comparisons between cages and more extensive systems then it
is essential that the data should be of equal validity. Again
the most effective way of doing this may be by concentrating
EEC resources of staff and expertise at a limited number of
centres, rather than dissipating them as we are tending to do
at present.

Those are some random ideas. I do hope one outcome of
our present meeting will be a move towards closer collaboration
between the animal welfare research programmes in our different
countries. For we are all in the same boat together now, in
that future regulations that may come from Brussels will apply
to us all. The logical response must be for us to work more
closely together, so that we can seek solutions in the most
effective way to the perfectly legitimate public concern with
animal welfare in an industry which must remain profitable to
remain viable.

CLOSING REMARKS

R. Moss *(UK)*

Before we close the meeting I would like to thank, first of all, the people who have presented papers. I thank them on my own behalf for having responded so well to my letter of invitation and also on behalf of the Commission.

I thank our three chairmen, Dr. Signoret, Dr. Adler and Professor Raymond, who have conducted our Sessions so admirably. You probably realise now why I asked Professor Raymond to finish the meeting; it was because I knew there would be a lot of stimulating ideas in his remarks.

May I thank my other colleagues from the member states for having come and for having contributed so well and so formidably.

I would like to thank Dr. Fischbach and Dr. Wagner for their help in arranging accommodation and meeting us, and Mr. Connell for his help in arranging administrative matters.

Finally, I would like also to thank Mick Hallam and Molly Robins of Janssen Services for dealing so admirably with the recording of every part of the meeting.

May I wish you a safe journey home.

LIST OF PARTICIPANTS

Prof. Dr. H.C. ADLER
    Royal Veterinary and Agricultural
    College
    Bülowsvej 13
    1870 Copenhagen V
    Denmark

Dr. W. BESSEI
    Universität Hohenheim
    Postfach 106
    7000 Stuttgart 70
    Federal Republic of Germany

Dr. G. BEUVING
    Instituut voor Pluimveeonderzoek
    "Het Spelderholt"
    Spelderholt 9
    7361 DA Beekbergen
    The Netherlands

Dr. J.M. BIENFAIT
    Faculté de Médecine Vétérinaire
    rue des Vétérinaires 45
    1070 Brussels
    Belgium

Dr. G.C. BRANTAS
    Instituut voor Pluimveeonderzoek
    "Het Spelderholt"
    Spelderholt 9
    7361 DA Beekbergen
    The Netherlands

Mr. G.J. BRESLIN
    Commission of the European Communities
    DG XIII
    Batiment Jean Monnet
    Kirchberg
    Luxembourg

Dr. G.M. CHIERICATO
    Istituto di Zootecnica
    University of Padua
    Via Gradenigo 6
    35100 Padua
    Italy

Mr. J. CONNELL
    Commission of the European Communities
    DG VI
    86 rue de la Loi
    1040 Brussels
    Belgium

Dr. I.J.H. DUNCAN
    ARC Poultry Research Centre
    West Mains Road
    Edinburgh EH9 3JS
    United Kingdom

Mr. J.M. FAURE	ARC Poultry Research Centre West Mains Road Edinburgh EH9 3JS United Kingdom
Ing. V. FISCHBACH	Administration des Services Techniques de l'Agriculture 16, route d'Esch - B.P. No.1094 Luxembourg
Dr. D.W. FÖLSCH	Institut für Tierproduktion ETH-Zentrum 8092 Zurich Switzerland
Dr. R. FRISCH	Administration des Services Techniques de l'Agriculture 3, rue de Strasbourg Luxembourg
Prof. J. FRIS JENSEN	Royal Veterinary and Agricultural College Bülowsvej 13 1870 Copenhagen V Denmark
Dr. W. GOLDHORN	Commission of the European Communities 86 rue de la Loi 1040 Brussels Belgium
Mr. S.E.W. HALLAM	Janssen Services 14 The Quay Lower Thames Street London EC3 United Kingdom
Mr. C.M. HANN	MAFF Great Westminster House Horseferry Road London SW1P 2AE United Kingdom
Dr. J.A. HILL	MAFF Experimental Husbandry Farm Meden Vale Mansfield, Gleadthorpe Nottinghamshire United Kingdom
Dr. D.E. HOOD	Meat Research Dept. Dunsinea Research Centre Castleknock Co. Dublin Ireland

Dr. B.O. HUGHES

ARC Poultry Research Centre
West Mains Road
Edinburgh EH9 3JS
United Kingdom

Ir. L.H. HUISMAN

Proefstation voor de Rundveehouderij
Rynderweg 6
8219 PK Lelystad
The Netherlands

Prof. P. LEYHAUSEN

Max Planck Institut für Tierhaltens-
physiologie
Wuppertal - Elberfeld
Federal Republic of Germany

Dr. G. MARTIN

IM Wolfer 56
7000 Stuttgart 70
Federal Republic of Germany

Mr. P.J. McARDLE

Poultry Division
Department of Agriculture
Agriculture House
Kildare Street, Dublin 2
Ireland

Dr. R. MOSS

MAFF
Government Buildings
Hook Rise South
Tolworth, Surrey
United Kingdom

Mr. R. NAGEL

Commission of the European Communities
200 rue de la Loi
1049 Brussels
Belgium

Dr. B. NICKS

Faculté de Médecine Vétérinaire
45, rue des Vétérinaires
1070 Brussels
Belgium

Prof. Dr. J. PETERSEN

Institut für Tierzucht und
Tierfütterung
University of Bonn
Endericher Allee 15
5300 Bonn
Federal Republic of Germany

Mr. M. PRIP

Royal Veterinary and Agricultural
College
Bülowsvej 13
1870 Copenhagen V
Denmark

Prof. W.F. RAYMOND

MAFF
Great Westminster House
Horseferry Road
London SW1P 2AE
United Kingdom

Mrs. M. ROBINS

Janssen Services
14 The Quay
Lower Thames Street
London EC3
United Kingdom

Dr. A. SCHILTGES

Administration des Services Techniques
de l'Agriculture
3, rue de Strasbourg
Luxembourg

Mr. P.M. SCHENK

Vakgroep Pluimveeteelt
ZODIAC, Postbus 338
6700 AM Wageningen
The Netherlands

Dr. J.P. SIGNORET

INRA
Station de Physiologie de la
Reproduction
Centre de Recherches de Tours
37380 Nouzilly
France

Dr. L. SPANOGHE

Faculteit van de Diergeneeskunde
Casinoplein 24
9000 Gent
Belgium

Dr. W. SYBESMA

Driebergseweg 10d
I.V.O. "Schoonoord"
Zeist
The Netherlands

Dr. R. TAUSON

Swedish University of Agriculture
Sciences
Dept. of Animal Husbandry
Poultry Division
Funbo-Lövsta
755 90 Uppsala
Sweden

Dr. K. VESTERGAARD

Royal Veterinary and Agricultural
College
Bülowsvej 13
1870 Copenhagen V
Denmark

Dr. E. WAGNER

Administration des Services Techniques
de l'Agriculture
16, route d'Esch - B.P. No.1094
Luxembourg

Prof. R.-M. WEGNER

Institut für Kleintierzucht der
Forschungsanstalt für Landwirtschaft
Dörnbergstrasse 25/27
31 Celle
Federal Republic of Germany

Prof. M. ZANFORLIN

Istituto di Psicologia
University of Padua
Piazza Capitaniato 3
35100 Padua
Italy